EMERGENCY LIGHTING

LIGHTING

&

POWER PROJECTS

No. 1788
$18.95

EMERGENCY LIGHTING
&
POWER PROJECTS

BY RUDOLF F. GRAF & CALVIN R. GRAF

TAB BOOKS Inc.

BLUE RIDGE SUMMIT, PA. 17214

FIRST EDITION

FIRST PRINTING

Library of Congress Cataloging in Publication Data

Graf, Rudolf F.
    Emergency lighting and power projects.

    Includes index.
    1. Emergency power supply.   2. Emergency lighting.
I. Graf, Calvin R.   II. Title.
TK1005.G68   1984   621.31'2   84-8865
ISBN 0-8306-0788-9
ISBN 0-8306-1788-4 (pbk.)

Photographs by Rudolf F. Graf and Calvin R. Graf unless otherwise indicated.

Cover illustration by Larry Selman.

Contents

And God said, Let there be light: and there was light.

(Genesis 1:3)

Introduction

Because modern man has become used to the wonderful, bright light of his lit-up world, he is thrown into a cultural shock when the lights go out, especially at night or in a closed-to-the-outside-world building or apartment. In this book, we cover the gamut of devices that produce electricity and generate light for use in our everyday lives, but especially when we experience a power outage. Covered in detail are topics such as home and intruder lighting, flasher lights and flashlights, home and camp power generation, batteries and chargers, security lighting and safety, and battery tester accessories.

The book is liberally illustrated with photos, schematic diagrams, and tables of electrical devices that are readily available at local stores. Several do-it-yourself projects and ideas are discussed so that the home electronic hobbyist can build these easy-to-construct items. They include emergency battery-operated safety lights, blinkers, flashers, piezosounder alerting devices, and home and camp power generators for use during power outages, camping and hunting trips, or at the farm or retreat lodge.

There is something for everyone in this book—the home owner, camper, hunter, jogger, experimenter, and innovator—interested in emergency lighting and power generation for home, camp, and security.

Everyday
1 Lighting and
Your Energy Dollar

This book covers many of the aspects associated with emergency lighting and power generation as might be used around the home, farm, campground, or automobile for safety and security. Electrical power generation for lighting, appliance operation, security, and safety are discussed at length. In order to have a better understanding and appreciation of electrical lighting and power and its conservation for use during emergencies and recreation we will first look at some aspects of familiar everyday light sources, their efficiencies, and color rendition.

LIGHTING

Modern civilization has enjoyed living with an abundance of electrical power. We grew up with it and have always expected it to be available in adequate quantities. It has only been within the past decade, since the Middle East oil embargo, that we have really become energy conscious. Since then, electricity users, both consumers and industry, have become acutely aware of electrical energy conservation. We have learned that if we use less electricity, we can save money. One of the ways we can use less electricity is to watch our lighting habits.

About 20 percent of all electricity generated in the United States today is used for lighting homes and businesses. In the Sunbelt region of the United States, the air conditioning portion of the electricity bill in the summer months may be as large as 50 percent, the rest going for lighting, utilities, and accessories. Of that portion spent for lighting, between 20 and 50 percent could be conserved by any home or business owner, with little capital expense, if any.

Energy conservation in lighting can be achieved with little effort on our part, sometimes as easily as flipping off a light switch when the light is not needed. Additionally, curbing energy use in lighting does not require a change in our lifestyle or a substantial investment of funds.

As a result of decades of use, consumers have come to rely on the familiar incandescent bulb for all of their lighting needs, even though different tasks may require different lighting types and levels—from high-powered outdoor security lights of hundreds of watts to very dim night lights as small as 4 watts. A variety of light sources is available, each of them having characteristics that make them suitable for different lighting tasks. This section will be helpful to a consumer who wants to select the most energy-efficient light source for his needs, whether he is simply replacing a bulb or remodeling a whole lighting system.

There are two considerations that are most important in choosing home lighting—energy efficiency and color rendition. We will examine the most commonly used indoor and outdoor lighting sources and ways to increase their operating efficiency. Such simple considerations as turning off lights when they are not needed, replacing bulbs as they dim, and keeping fixtures clean have a surprising beneficial impact on energy usage and the monthly electric bill. Where light intensity is not important, replace larger wattage bulbs with small wattage ones.

If you will be out of a room for five minutes, turn the light off. Turn the TV set off when no one is watching. When you just want background music, turn on a small transistor radio instead of the larger stereo set. All of these actions will save electrical energy and help make your electric bill smaller each month. Pay yourself instead of the power company.

Energy Efficiency

Energy efficiency for all light sources is the measure, or ratio, of how much light is produced—measured in *lumens*—in relation to the amount of energy used—measured in *watts*. Lumens-per-watt ratings are similar to a very familiar measure of energy efficiency for an automobile—miles per gallon. Just as cars have different efficiency ratings, so do light sources or "lamps." Some lamps convert electricity into light much more efficiently than others, and can, therefore, deliver more light for the same amount of electricity.

For example, a 40-watt fluorescent tube that delivers 66 lumens per watt is more than five times as efficient as a 40-watt incandescent bulb that delivers only 12 lumens per watt. The incandescent lamp is inefficient because most of its energy lies in the heat frequency range (infrared) and not the light frequency range which the human eye can see. The difference in the amount of light provided for each watt can have a dramatic impact on the amount of electricity required to light a house or commercial building. Likewise, when we are out camping, the more light sources we can have for a given amount of electrical generating capability, or the longer our gasoline supply will last before we have to refill the portable generator the more cost-effective our camping will be.

New, highly efficient lamps are available that produce considerably more light for every watt of electricity than older, more familiar light sources. By using these energy-efficient lamps, a homeowner can reduce energy consumption while maintaining desired lighting levels. The efficiency of different light sources varies considerably—from less than 10 lumens per watt (LPW), for some incandescent bulbs, to over 130 lumens per watt for high-pressure sodium lamps. The range in efficiency of these light sources is shown in Fig. 1-1.

Color Rendition

Energy efficiency is one of the factors a homeowner or builder considers in choosing a lighting system, especially when building a new home or modernizing his present installation. And of course, he also wants the light source to create a pleasant atmosphere. Color rendition is an important concern in lighting aesthetics, but it is a characteristic of light sources that is difficult to evaluate. Color is the effect of light waves from a source bouncing off or passing through various objects until it reaches our eyes. The color of a given object is, therefore, determined in part by the characteristics of the light source under which it is viewed.

Color rendition is a relative term. It refers to the extent to which the perceived color of an object under a light source matches the perceived color of that object under the familiar incandescent bulb. To say that a light source has a good color rendition can be translated as saying that it gives objects a familiar appearance—the appearance they would have under incandescent lighting.

The color-rendering properties of incandescent bulbs are the

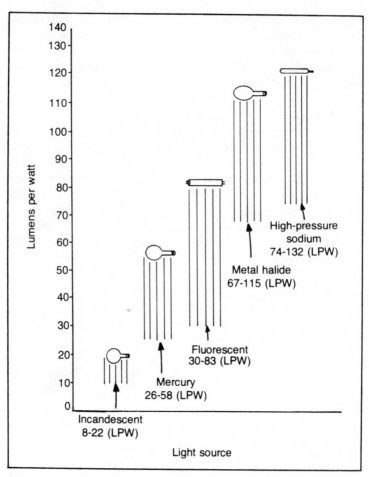

Fig. 1-1. Light efficiency versus light source.

standard measurement for indoor lighting because they are the most commonly used light source, and people have become accustomed to the appearance of objects under them. The incandescent lamp produces light which is most like that from the sun, which we usually call "white light."

In bright sunlight, a red apple is red and green leaves are green, as is true when we see them under light from an incandescent lamp. If we look at the red apple and green leaves under the red light we find in a darkroom, we would see the apple as still being red, but the green leaves would now appear to be black. Under mercury vapor lights such as are used for street lighting or home security lighting, skin tones are distorted and a girl's red lipstick looks deathly dark. In this case, the color rendition is not true!

The incandescent lamp accentuates warm tones—reds, oranges, yellows—while the standard fluorescent lamp used in most offices and stores accentuates cool tones—blues and greens. For this reason, people have been reluctant to use fluorescent lighting in their homes.

EVALUATING LIGHT SOURCES FOR HOME AND CAMP

Residential and camp lighting must balance efficiency and aesthetics while being adaptive to the wide variety of activities within the home and camp. And most desirable, the lighting must be economical. In this section, we will see how well these factors are balanced for the three major light sources—incandescent, fluorescent, and high intensity discharge lamps used for outdoor lighting.

Incandescent Lighting

Incandescent lamps are by far the most widely used source of lighting in homes and camps throughout the United States and most of the world where electricity is available. They are also by far the most energy wasteful, as almost 90 percent of the electrical energy consumed by the incandescent lamp is dissipated as heat, not light energy. This is due to the fact that most of the energy output from the incandescent lamp lies below the light region, in the infrared (heat) portion of the electromagnetic spectrum and only about 10 percent of it extends up into the visible region of the human eye. Incandescents waste so much electricity as heat because they create light by an intermediary process. They use electricity to heat a coiled tungsten filament in a vacuum bulb until the filament glows, rather than converting electricity directly into light as does the light emitting diode (LED). The light output of the LED is presently not to the point where it will be a threat to the incandescent lamp but research into its applications for lighting purposes makes the future of solid-state light bright.

Efficient Use of Incandescents. Even though incandescent lighting is not energy efficient, efficiencies can be achieved by choosing the right type bulb and wattage size to fit the desired application. The first thing to know is how to read and compare the information printed on the package.

An important rule of thumb to remember in determining efficiencies of incandescents is that the efficiencies increase as the wattage increases. For example, one 100-watt incandescent bulb has an output of 18 lumens per watt, while a 60-watt bulb produces only 14 lumens per watt. So if we substitute one 100-watt bulb (1800 lumens) for two 60-watt bulbs (1680 lumens), it produces more light and uses less electricity. This type of substitution saves energy and reduces maintenance, and therefore should be made whenever possible.

Life Span of Incandescent Lamps. The life span of a bulb has significant impact on the efficiency of incandescent lamps. The average life span of a bulb and its light output in lumens is printed on the package. Table 1-1 shows the wattage, lumen output of the lamp, and average life span of the most common wattage sizes in use in the United States where 110-120 Vac power is generally available. Not only does an incandescent lamp have the shortest life span of all available lamps, but, near the end of its lifetime, the light output will also drop as low as 50 percent of its original output while still consuming the same amount of energy.

Light output from the bulb reduces over time because as the coiled tungsten filament in incandescent bulbs emits light, molecules of the metal burn off. These molecules are deposited on the inside of the glass bulb and slowly darken the bulb. As the bulb darkens, it consumes the same amount of energy as it did when it was new and yet it produces less light. The bulb eventually burns out when the filament ruptures. Almost all lamps appear to burn out with a brilliant flash when first turned on because at this time the filament is cold and the inrush current may be more than 10 times the operating current. Energy and money can be saved by replacing darkened bulbs before they burn out.

Long-life bulbs, which have a life span of 2500 to 3500 hours, are the least efficient incandescents because light output is sacrificed in favor of long life. Nevertheless, they are the best to use where light intensity or color rendition are not important, such as exit signs, night lights for hall safety, or driveway background lighting. They are especially useful where bulbs are difficult to replace such as high ceilings in homes, buildings, and churches.

When shopping for bulbs, look for those that provide the most lumens of light output for a given size bulb. A comparative reading of light bulb packages reveals that a tinted bulb has a lower light output than a standard incandescent bulb of the same wattage. This is because the coating on the bulb inhibits the transmission of light. The higher prices of tinted bulbs further illustrate that energy efficiency is often less expensive from the start. The cost of light sources is nearly always a minor consideration compared to the cost

6

Table 1-1. Life Span of Various Incandescent Lamps.

	Bulb Size Watts				
	25	40	60	75	100
Average Light Output Lumens	230	460	890	1210	1750
Average Operating Hours	2500	1500	1000	750	750
Lumens per Watt	9.2	11.5	14.8	16.1	17.5

of the power to light them. Under normal circumstances, the lamps will represent less than 10 percent of the cost of the lamps and energy combined. Therefore, the efficiency of the lamp is more important than its price. To achieve energy efficiency using incandescent bulbs, the homeowner should do comparative shopping, checking for lumen output, and bulb lifetime in hours, all of which are printed on the bulb package.

With all the knowledge we now have, it is easy to talk about the shortcomings of the incandescent lamp. But we should remember that until Thomas A. Edison invented the light bulb in 1879 man generally lived in darkness at night. Edison tried thousands of types of materials as a lamp filament. Some of them would burn for just seconds, some as short as a photo flash bulb. But on October 19, 1879, after many failures, Edison finally succeeded in placing a filament of carbonized thread in a bulb. This bulb shed good light and the precious bulb glowed for three days until Edison decided to increase the voltage. Not until then did the bulb burn out, but Edison's invention of the incandescent lamp astounded the world—and you can buy this wonderful invention for as little as twenty-five cents!

Tungsten Halogen Lamps. Attempts have been made to increase the efficiency of incandescent lighting while maintaining good color rendition. This work has led to the manufacture of a number of energy-saving incandescent lamps for use in residential applications. The tungsten halogen lamp varies from the standard incandescent lamp because it has halogen gases in the bulb. Halogen gases keep the glass bulb from darkening by preventing the filament

7

from evaporating and thereby increase the lifetime of the lamp to up to four times that of a standard bulb. The lumen-per-watt rating is approximately the same for both types of incandescents, but tungsten halogen lamps average 94 percent efficiency throughout their extended lifetime. This offers significant energy and operating cost savings. Remember, blackening of the bulb in a regular lamp may drop the light output to 50 percent as the bulb ages.

Tungsten halogen lamps require special fixtures, and during operation, the surface of the bulb reaches very high temperatures, so these lamps are not commonly used in the home. They have become very popular in the automotive field, however, and are frequently found mounted on pick-up trucks, campers, or vans, as original or optional equipment.

Reflector or R-Lamps. Reflector lamps(R-lamps) are incandescents with an interior coating of aluminum that directs the light to the front of the bulb. Certain incandescent light fixtures, such as recessed or directional fixtures, trap a portion of the light inside the fixture. Reflector lamps project a cone of light out of the fixture and into the room, so that more light is delivered where it is needed. These fixtures are very efficient because a 50-watt reflector bulb will provide better lighting and use less energy when substituted for a 100-watt standard incandescent bulb.

Reflector lamps are an appropriate choice for task or spot lighting, because they directly illuminate a work area, and are excellent for accent lighting. Reflector lamps are available in 25, 30, 50, 75, and 150-watt sizes. While they have a lower initial efficiency (10 lumens per watt) than regular incandescents, they direct light more effectively, so that more light is actually delivered than with regular incandescents.

PAR Lamps. Parabolic aluminized reflector (PAR) lamps are reflector lamps with a lens of heavy durable glass, which makes them an appropriate choice for outdoor flood and spotlighting. PARs have been popular for a number of years when used as outdoor floodlighting since they are available in 75, 150, and 250-watt sizes. Even though they are a higher wattage bulb, they have longer lifespans with less depreciation than standard incandescents.

ER Lamps. Ellipsoidal reflector (ER) lamps are ideally suited for recessed fixtures because the beam of light that is produced is focused 2 inches ahead of the lamp to reduce the amount of light that is trapped in the fixture. In a directional fixture such as the ER lamp, a 75-watt ellipsoidal reflector delivers more light than a 150-watt R-lamp. The ER lamp produces more useful light because as a reflector lamp it delivers more light than standard incandescents in directional fixtures.

Fluorescent Lighting

Fluorescent lamps do not depend on the buildup of heat for light as do incandescent lamps. Rather, fluorescents convert energy to light by using an electric charge to excite gaseous atoms within the fluorescent tube, very similar to the action of the laser light generator. The charge is sparked in the ballast and flows through cathodes in either end of the tube. The resulting gaseous discharge produces ultraviolet light which is not visible. The ultraviolet radiation energizes the phosphor coating on the inside wall to *fluorescence* and it radiates strong light in the visible spectrum. Because the buildup of heat is not requisite to the creation of the light, the energy wasted as heat is significantly less than is wasted by incandescent lighting. The process by which fluorescent lamps convert electricity to light is up to five times as efficient as the incandescent process.

Fluorescent lamps require a special fixture for operation. For the standard, straight fluorescent tube, this fixture consists of a metal channel that contains the *ballast*, a small transformer that regulates and limits the flow of current through the tube. The fixture will have a small round device about the size of a spool of thread called the *starter* that aids in starting the lamp. Rapid-start lamps are also available. The start-up delay associated with fluorescent lamps is very brief. The lamp will flicker for a few moments when first turned on before achieving its full brilliance. There are also U-shaped and circular fluorescent lamps that come complete with adapters which eliminate the need for special fixtures and wiring modifications.

Color Rendition with Fluorescent Lighting. The color-rendering properties of fluorescent bulbs are determined by a coating applied to the inside of the bulb. The bulbs are labeled "warm" or "cool" to indicate the effect their use creates. You can choose among four kinds of commonly used fluorescent bulbs, including cool white, deluxe cool white, warm white, and deluxe warm. Each type has different color rendering properties and different levels of efficiency.

Fluorescent Lights and Efficiency. You will save a lot of energy when you use a fluorescent lamp because efficiency and lamplife are so great compared to an incandescent lamp. The energy savings are not due simply to the fact that you are using a smaller lamp but also to the fact that a fluorescent lamp will operate more efficiently over a longer period of time. The efficiency of a fluorescent lamp will increase as the length of the lamp tube is increased. It follows that whenever possible you should use the larger lamp fixture as you will save energy.

The ballast in the fluorescent fixture consumes a small but constant amount of energy, even when a tube has been removed. So when removing a tube out of a series in order to conserve power, you will want to disconnect the ballast or unplug the fixture entirely when it is not in use. One ballast in a string fixture is often used by more than one fluorescent lamp, so removing one lamp will cause the others to go out also. A ballast that hums loudly may be ready to go out. If it is not replaced in a short while, it could heat up sufficiently to cause the ballast to smoke. The extra current drawn by the smoking ballast may not be great enough to trip the associated circuit breaker or fuse. Proper maintenance of the fixture is, therefore, very important.

Switching Lights Off When Not in Use. About a decade ago, when electricity was less expensive, many owners of large buildings left their fluorescent lights on all the time. Maintenance costs, due to switching them on and off, were greater than electricity costs. While most everyone switched off power-hungry incandescent lamps when not in use, there was a lot of misunderstanding about what to do with fluorescent lamps. Should they be turned on and off? We know now, however, that fluorescent lamps, just like incandescents, should be turned off when not in use, even if only for a few minutes. No energy is required to turn a light off, and the initial charge required to turn a fluorescent lamp back on does not use a significant amount of energy unless the switch is flipped on and off very rapidly.

Fluorescent lamp life is rated according to the number of hours of operation for each start, and while it was once true that the greater number of hours operated per start, the longer the lamp life, recent advances in technology have increased lamp life ratings to an extent that makes the number of starts much less important than they were 10 or 20 years ago. Therefore, as a rule, if an area is to be unoccupied for more than 15 minutes, or so, fluorescent lamps should be turned off.

WAYS TO IMPROVE LIGHTING EFFICIENCY

There are a number of ways to improve lighting efficiency so that you will get the most illumination for the least cost. These involve correct lamp placement, regular maintenance, and efficient operation. The use of timing and dimming devices also contributes to increased lighting efficiency and energy savings and will be discussed in Chapter 3.

Placement

Efficiency of light fixtures can be significantly improved by placing them properly in relation to the objects that surround the fixtures and the activities to be illuminated. For example, two lamps that are placed incorrectly may be needed to adequately illuminate a work area where only one properly placed lamp would actually be required. Important factors to be considered:

- Light-colored walls and bright surfaces reflect more light than do dark surfaces. These rooms will also appear to be larger than darker-colored rooms. Light-colored carpets also make the room appear lighter, brighter, and larger.
- Perimeter lighting or lamps spread around the room usually give more brightness than does central lighting.
- A directional lamp used above a work area gives more illumination for the same wattage bulb than does more diffused lighting.
- The balance of lighting in a room should be evenly maintained. This is especially so in a room having both general lighting and task lighting over work areas such as kitchen counters or desks. Balance should be maintained where the two types of lighting meet. If the various sources of light do not reach far enough to meet, you may have excessive shadows and contrast areas between the light and dark areas and this may cause eye fatigue. Adding another lamp or increasing the wattage of the existing ones, however, is both expensive and wastes energy. To solve this properly, lamps should be repositioned to spread out the light and distribute it more evenly. Periodically look at the way in which rooms are used to determine whether the efficiency of lighting can be increased.

Maintenance

The efficiency of a light source depends to a large extent on how well the fixture is maintained, be it house lighting, camp lights, or flashlights. A lamp that produces about 18 lumens per watt (a 100-watt bulb) when installed, may actually distribute only 19 lumens when covered with dust. Clean bulbs regularly in both lamps and ceiling fixtures, for they collect dust in an air-conditioned house or building even though filters ar installed. The reflecting value of fluorescent fixtures is important to their illuminating ability, and their tendency to collect dust can be coped with by frequent dusting and cleaning with a damp cloth.

Incandescent bulbs should be replaced when they first begin to appear dim or when they begin to blacken; these are signs that they will burn out shortly. Remember, a bulb seldom burns out when it is in use. It is most likely to pop when you first turn it on—often at a most inappropriate time, such as when you are in a hurry to get to work. Lamp dimmers will help this situation as will be discussed later.

Operation

You will be able to affect significant energy savings simply by using light sources efficiently. The following are some suggestions which will help increase lighting efficiency:

- Use lower wattage bulbs where possible. Rely on daylight where you can.
- Replace two bulbs with one having a comparable number of lumens where you can. But remember, ceiling fixtures may have a maximum of 60-watts per lamp, so do not use a 75 or 100-watt bulb to get more light. They generate more heat close to the ceiling.
- For night light operation, substitute a 4-watt bulb for the standard 7-watt bulb. These 4-watt bulbs have a clear glass finish, are almost as bright as the 7-watt bulb, and yet use only about half as much energy.
- To save power, place fixtures on separate switches so they can be operated independently. You can turn on only those lights you need, and not others that do not contribute to your work area.
- Decorative outdoor gas lights use a lot of energy because they are on for 24 hours a day. Convert them to incandescent lamps where feasible and use a light sensor to turn them on at dusk and off at dawn. We will discuss the use of light sensors, dimmers, and diodes in Chapter 3.

Reference: Lighting, U.S. Department of Energy, DOE/CS-000-, January 1978.

2 | Emergency Lighting

Emergency lighting means different things to different people at different times. When an electrical power failure occurs in the daytime and you are at home in your house in the suburbs, the loss of power means little because you will have some natural light helping you to find a flashlight or a candle and matches, and the transistor radio. If you are in a large building when the power goes off, the whole building dies electrically, and you are entirely dependent upon installed emergency lighting that you hope will operate properly. If you are in an area without a window, such as storerooms and hallways, you will be thrown into pitch darkness. You will be dependent upon battery-supplied lighting until the power comes back up and the lights come on.

A young child asleep in his bedroom when the power goes off could be extremely frightened if he awakens to find himself in a pitch black room. He would have no frame of reference and he might panic. A small battery-powered night light, adjusted to come on when the power goes off, could allow that child to see familiar surroundings and feel safe.

Place yourself in a movie theater when the whole audience is thrown into darkness by a power failure. Without emergency lights to provide a frame of reference, the whole crowd might panic as they stumble toward an exit which they can only imagine is somewhere. Just one small glowing light, like that from a match or lighter or battery-powered light, can help calm people's fears. Fortunately,

13

most states have building codes which require emergency lighting and exit signs in strategic locations in public buildings, large private buildings, restaurants, and theaters, and in hotel or motel halls, stairwells, and lobbies. They will function when needed, however, only as well as the maintenance performed on them.

POWER FAILURE AND EMERGENCY LIGHTING

Let us now discuss various battery-powered emergency lighting devices available on the open market. The items covered will help you decide what best fits your needs so that you will be prepared for any eventuality.

Rechargeable Security Light

The rechargeable Dark Blazer Model PF2 is a flashlight which is left plugged into any 120-Vac outlet and turns on automatically when there is a power failure. Figure 2-1 shows the PF2 in a display box. When the PF2 is plugged into a hall or room outlet, a red light-emitting diode (LED) serves as a night light and also indicates the unit is charging. The unit is kept continuously charged in this manner until a power failure. When there is a failure of just a few seconds, the power failure light comes on and the LED goes out. If you want to use the light, you simply unplug it from the wall outlet

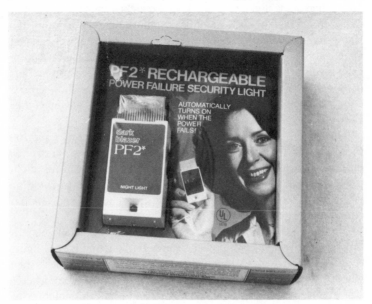

Fig. 2-1. Dark Blazer PF2 rechargeable flashlight in display box.

Fig. 2-2. Dark Blazer PF3 with audio alert feature.

and use it as a compact flashlight. It has a concentrated beam for distant objects but you can also use it as a full-circle light; it will provide you with 90 minutes of dependable light for emergencies or essential activities. The light will eliminate dripping candles and open flames.

A switch allows you to turn it on and off like a regular flashlight so that when you plug it back into the wall outlet after the power has come back on it will recharge automatically until your next occasion to use it. The flashlight bulb replacement is easy and can be done in less than a minute.

The PF2 power failure light is available from local stores and is manufactured by Nicholl Brothers, Incorporated, 1204 West 27th Street, Kansas City, Missouri 64108.

Audio Alert Security Light

The Dark Blazer Audio Alert Model PF3 is also made by Nicholl Brothers, Incorporated, and is shown in Fig. 2-2. This unit

Table 2-1. Specifications for PF3 Audio Alert Rechargeable Flashlight.

Lens Design	Concentrated beam w/360 degree area light
Audio Alert	75 dB solid-state unit
Battery System	90 minute capacity—NiCad
Bulb	#14 standard
Night Light	Light-Emitting Diode (LED)
Power Consumption	0.4 watts
Size (w/o Prongs)	3.7 h × 1.85 w × 1.54 inches
Weight	3.6 ounces
Case	High-impact material

is similar to the Model PF2 but it also includes a power-failure audio alert. When switched on and there is a power failure, the audio alert sounds a compelling alarm. If you are asleep when the audio alarm goes off, you will be awakened and can then take any required actions. Many electronic digital alarm clocks lose the time-keeping function when there is a power outage even as short as several seconds. So if your digital clock does not have a battery backup to keep the time clock going, you will not be awakened on time. When you are awakened by the PF3 audio alarm, you can set a mechanical alarm clock, go see how the children are doing, check the doors of the house or apartment, or reset the electric clock.

In addition to the alarm, the rechargeable flashlight also comes on so that you can easily locate the unit to turn the audio alarm off. You can unplug the unit and use it as a flashlight with its own on-off switch. One switch turns the alarm on and off, when unplugged from the wall, and one switch turns the flashlight on and off, so you can conserve the power until you need it. The PF2 would probably be the best unit to place in a young child's room because it can serve as a night light without disturbing the child's sleep. The PF3, however, with its audio alarm, would probably awaken the child and possibly frighten it. The PF3 is ideal for the parents' bedroom. Table 2-1 provides details on the Model PF3 unit.

The PF3 Audio Alert Security Light is manufactured by Nicholl Brothers, Inc., 1204 W. 27th Street, Kansas City, Missouri 64108 and is available from local stores.

Power Failure Light

Radio Shack sells the Archer power failure light shown in Fig. 2-3 as Catalog No. 61-2645. This light uses nickel cadmium batteries to provide up to 90 minutes of light during a power outage. The unit also plugs into a wall outlet for recharging and can be used as an ordinary flashlight by simply unplugging it from the wall. An LED always glows when the unit is plugged in so it serves as a convenient night light.

Sound-Activated Emergency Light

A novel smoke-penetrating emergency light manufactured by Jameson Home Products, Incorporated is shown in Fig. 2-4. Code

Fig. 2-3. Archer Power Failure Light (courtesy of Radio Shack, a division of Tandy Corporation).

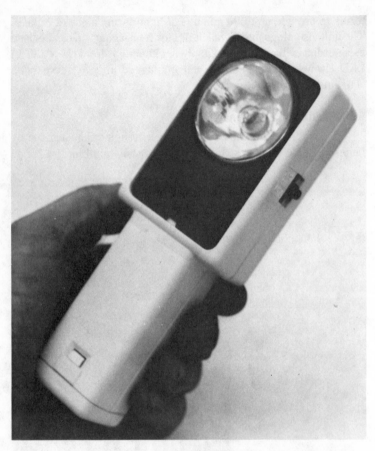

Fig. 2-4. Code One Emergency Light is activated by any smoke alarm.

One Emergency Light Model No. CD-30 is activated by any nearby ringing smoke alarm. When a smoke alarm sounds, you can locate the flashlight, remove it from its holder and carry it to locate family members and help light their escape path. Detailed operating instructions are provided with the unit so that you can install it properly.

The Code One Emergency Light, which has a yellow beam to better penetrate smoke, will work very effectively in a normal home environment if it is mounted within 20 feet of your smoke alarm. If your smoke detector is approved by Underwriters Lab, it will have an 85 dB horn. The loud sound from the smoke alarm will turn on your emergency light. Locate a permanent mounting location for the Code One Light by holding the unit with its mounting bracket in a desired location on the wall. Have someone activate the smoke alarm. If the alarm turns the light on, you can proceed to permanently fasten the mounting bracket to the wall. If the light does not come on, move the unit closer to the smoke detector and repeat the test.

The alarm and light should be tested weekly to ensure proper operation. When the four AA alkaline batteries become weak, the Code One light will emit a low chirping sound at approximately one minute intervals. The unit is designed to emit this sound for up to a week should you be gone on a trip when the batteries begin to weaken. They should be replaced as soon as possible so that the light is ready to be used at any notice. If you have used the light as a flashlight in nonemergency situations for any period of time, the low-battery chirp may sound much earlier. The four alkaline batteries are installed in the handle of the flashlight as shown in Fig. 2-5.

The sound sensor for the Code One is designed to respond to the sound output of any UL-listed smoke alarm with an 85 dB horn. The light will respond only to very high sound levels which persist for approximately four seconds and fall within the smoke alarm

Fig. 2-5. Four alkaline batteries power Code One flashlight.

output frequency band of 2.2 kHz-3.8 kHz. Other frequencies are filtered out. Loud stereo amplifiers, for example, could not set off the light because the sound frequency would not persist for the necessary four seconds. The Code One light turns itself off approximately eight seconds after the smoke alarm ceases.

The Code One will also provide portable protection when you travel to motels and hotels equipped with smoke alarms. To operate the unit as a portable, slide the switch to Auto and place the light on any hard, flat surface such as a night stand or table so that the spring-loaded pin switch on the back of the alarm is depressed. The Code One light is now ready to respond to any UL-listed smoke alarm just as it would in it's wall mounting back home. The light will stay on whenever you switch to Auto and either remove the unit from its bracket or pick it up from a flat surface. You can turn the light off by laying the unit down again or switching to the Off position. Details on the Code One Emergency Light are shown in Table 2-2.

Power Failure Emergency Light

This unit is made by Snapit and is an extremely useful emergency light to indicate a power failure: when the power goes off, the light comes on! Shown in Fig. 2-6, the unit plugs into any wall socket. A built-in automatic neon light comes on when the unit is plugged into the wall outlet. The neon light serves as a handy night light, especially when placed in a hallway. However, when there is a power failure, a powerful beam of light comes on and is concentrated by a Fresnel lens. The unit can also be used as a handy

Table 2-2. Specifications for Code One Light.

Detection	Detects any UL-listed 85 dB smoke alarm horn within 20 feet (range varies)
Sensitivity	Discriminator detects 2.2 kHz-3.8 kHz passband; minimal 80 dB
Circuitry	100 percent solid-state
Operating Switch	Auto/Off/Test
Battery	Uses four AA 1.5 alkaline cells
Low Battery Indication	Beeps approx. once per minute; seven days minimum
Bulb	High intensity, replaceable PR13 bulb
Lens	Tinted yellow
Bracket	High impact styrene; includes two screws
Temperature Range	40° to 100° F.
Humidity Range	20 to 85 percent relative humidity
Weight	8 ounces

Fig. 2-6. Power Failure Light by Snapit.

flashlight by simply unplugging it from the wall outlet.

A test switch on top of the unit lets you test the bulb as well as the condition of the two replaceable AA alkaline batteries while it is plugged into the wall. Brightness of the light shows the condition of the batteries. Compact in size, the unit measures $2\frac{1}{2} \times 2\frac{1}{2} \times 2$ inches. New alkaline batteries should last about two years under normal conditions. Figure 2-7 shows the compartment for the two

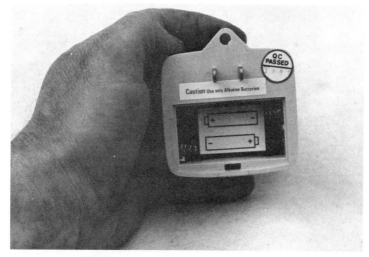

Fig. 2-7. Two alkaline AA cells power Snapit flashlight during power outage.

cells which have enough energy to provide light for about 1½ to 2 hours during a power failure. The unit is designed to hang on a peg hook or stand by itself. The fully solid-state unit is rated at 110-120-Vac, 50-60 Hz and draws 1 watt of power when plugged in.

The Snapit emergency light is manufactured by Cable Electric Products, Incorporated, P.O. Box 6767, Providence, Rhode Island 02940, and is available at stores throughout the country.

LIGHT-SENSITIVE DEVICES

Many commercial products now contain photocells (light-sensitive resistors) which allow night lights to turn on and off in response to existing light. Three products described here provide light only when you need it—when it is dark!

Sensor-Lite

The Sensor-Lite made by Snapit serves as a night light when it is dark. It shuts off automatically at dawn to conserve power. The Sensor-Lite uses a light-sensitive resistor (photocell) to sense the amount of light present; the solid-state circuitry gradually turns the 7-watt night light on as the evening approaches. This allows you to conserve energy whenever it is light enough near the Sensor-Lite. The light can be placed in garages, hallways, bedrooms, or kitchens, and provides inexpensive security when you are away from home. The Sensor-Lite uses the standard C-7C clear or frosted household bulb available at most hardware and utility stores. The unit is manufactured by Cable Electric Products, Incorporated, P.O. Box 6766, Providence, Rhode Island 02940.

Automatic Night Light

Most Radio Shack stores carry an automatic night light as Catalog No. 61-2646. This light also uses a light-sensitive resistor to determine that darkness is approaching and turns on the small 7-watt bulb.

Goodnight Light

This automatic night light is similar to those described earlier. It plugs into a 120-volt wall outlet and will turn the light on at dusk and off at dawn. These night lights are sensitive enough to sense when a shadow crosses the photocell and they will turn the light on for just an instant if someone darkens a lighted hallway or room.

The Goodnight Light is manufactured by Diablo Technologies, Incorporated, 2694 Bishop Drive, San Ramon, California 94583.

ELECTRONIC POWER ALARM

This unit is plugged into a wall outlet and will sound a one-second beep alarm whenever there is a power interruption lasting longer than 10 seconds. Packaged under the Snapit tradename as Catalog No. 48630, the unit is an all solid-state sensor that uses a 9-volt alkaline battery to power the electronics and sound the alarm. Whenever the power goes off for more than 10 seconds, the alarm sounds and the LED flashes so you can locate the unit to turn it off. The blinking LED and audio signal will start every 15 to 20 seconds when the battery supply voltage drops below 7½ volts, an indication to replace the alkaline battery. A new battery will power the unit for several years and will sound the alarm for a minimum of five days. This is a handy unit to have installed in the same wall outlet as your refrigerator or deep freeze to alert you to the fact that the power to that device has been interrupted. You can then take action to identify and solve the problem.

The electronic power alarm is manufactured by Cable Electric Products, Incorporated, P.O. Box 6767, Providence, Rhode Island 02940.

EMERGENCY LIGHTING SYSTEMS

In the United States, the National Electrical Code and the Life Safety Code set standards for lighting requirements in public buildings. In this section, we will briefly discuss portions of the standards that are of interest to us.

Municipal, state, federal, or any governmental agency having jurisdiction over an area can define emergency system requirements for that area. These systems should automatically supply light and/or power to critical areas during a power failure. The emergency systems should serve as a backup to any system intended to supply, distribute, and control power and illumination essential for safety of human life. The main ingredients here are power and illumination for safety of human life.

Emergency systems are generally installed where artificial illumination is required for safe exiting and for panic control in such buildings as hotels, motels, theaters, sports arenas, and health care facilities. These emergency systems may also provide power for other functions such as ventilation, fire detection and alarm systems, and elevators. Emergency lighting systems must be designed and installed so that the failure of any individual lighting element, such as the burning out of a light bulb, cannot leave in total darkness any space which requires emergency illumination. The rechargeable lighting units described earlier are devices which do just that—they prevent a room from being plunged into pitch-black darkness. Our discussion here will deal with home emergency lighting and does not cover the commercial aspects of emergency lighting and power requirements.

ELG-2 Gemini Self-Contained Emergency Lighting Unit

A high-power, high-light output emergency lighting unit is shown installed above the entrance to a home in Fig. 2-8. The ELG-2 Gemini is lightweight and compact and is used as a permanent installation in a portion of the house to provide maximum protection during emergency power outages.

The unit is mounted up and out of the way and comes on only when there is a power outage. The ELG-2 uses two tungsten wedge-base lamps to provide brilliant light beams to shine up and down a hall or large room in a house. Some construction and design features are as follows:

Fig. 2-8. Gemini Emergency Lighting Unit installed over front door of residence.

- Operation. The Gemini provides a minimum of 90 minutes of automatic emergency illumination upon loss of local utility power. The solid-state electronics provide automatic recharge after 12 hours of use. The electronics also prevent full discharge and over- and under-charging of the battery.
- Readiness Indicators. The Gemini test switch permits a check of the system's operation. A Ready light emits a constant signal, indicating the readiness state or operating function.
- Installation. The Gemini installs quickly and easily, facilitated by a separate mounting chassis, hinged housing, and plug-in battery and lamp terminals.
- Lamp Heads. Molded of off-white thermoplastic, the lamp heads are designed to house tungsten lamps. The Lexan lens made by General Electric turns 90 degrees for beam pattern adjustment. Internal wiring is concealed. The lights are equipped with swivels for full adjustability.
- Lamps. The unit uses two 3.6-watt tungsten wedge-base lamps.
- Battery. The 4-volt, sealed, spill-proof, rechargeable, maintenance-free lead-calcium battery operates over a wide temperature range.
- Circuit Design. Gemini advanced, solid-state electronic circuitry is equipped for 120-Vac input. A low-Vdc disconnect feature protects the battery from deep discharge. Over- and undercharging are prevented by precisely regulated float voltage (allows plus/minus 12 percent line voltage variation). All electronic components are mounted on a printed circuit board a shown in Fig. 2-9.
- Dimensions. The Gemini is 10⅜ inches long, 3⅜ inches deep, and 4½ inches high. Lamp heads add 3¾ inches to overall height. The light weighs 5½ pounds.
- Maximum Amperage. The maximum amperage at 120 volts is 0.58 amperes.
- UL/Listed. Gemini complies with NEL, Life Safety Code, and OSHA standards.

The ELG-2 is manufactured by Lithonia Emergency Lighting Products, 2528 Lantrac Court, Decatur, Georgia 30035. A check of the yellow pages of your telephone directory under hardware electrical fixtures or contractors should provide you with information on availability of the unit.

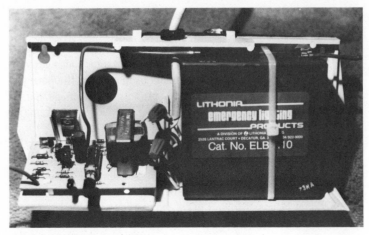

Fig. 2-9. Gemini ELG-2 with cover removed.

Mini Fluorescent Light

During an extended power outage, it is handy to have a light source which will provide illumination over a wider area than that covered by a small table flashlight. One such light is the Mini Fluorescent Light offered by Radio Shack as Catalog No. 61-2733. Shown in Fig. 2-10, this unit provides instant light for use at home or away at camp. The unit is surprisingly bright for its small size and is perfect for camping, fishing, or emergencies. It is small enough to fit conveniently in a glove compartment or your backpack. A carrying case attaches to your belt for easy carrying. The unit measures 6¼ × 3⅜ × 1¼ inches and requires five C cells. The unit is available at most local Radio Shack stores.

Compact Fluorescent Lite

This handy light uses a 6-inch, 4-watt fluorescent tube rated at 6000 hours. We see from Fig. 2-11 that it is similar to the Mini Fluorescent Light described above. The Compact Lite provides a brilliant, soft light. The unit uses five C cells and a transistorized circuit to boost the battery voltage to a value which will operate the fluorescent tube and yet provide long battery life. The housing around the tube is made of break-resistant clear plastic. The light is handy for home, auto, camper, or boat, or any light emergency. The Compact Fluorescent Lite is manufactured by Wonder Corporation of America, Norwalk, Connecticut 06856.

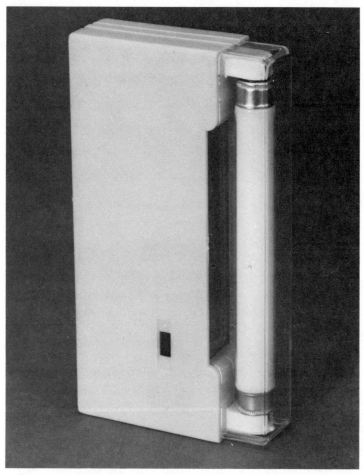

Fig. 2-10. Mini Fluorescent Light uses five C cells for power (courtesy Radio Shack, a division of Tandy Corporation).

Mini Fluorescent Light

This fluorescent light and flashlight is perfect for camping, highway emergencies, and blackouts. Figure 2-12 shows the two-function light which provides either a flashlight or a general-area light in a single unit. The light is small enough for a backpack or glove compartment and has a handy carrying cord. The unit measures 8 × 2¼ × 1⅜ inches and requires three C cells to operate.

The Mini Fluorescent Light is manufactured by Radio Shack and is carried as Catalog No. 61-2734.

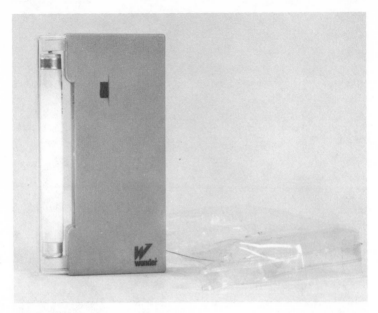

Fig. 2-11. Compact Fluorescent Lite similar to that of Fig. 2-10 (courtesy Wonder Corporation of America).

EMERGENCY LIGHT-SOURCE PROJECT

There are many emergency light sources in use today. Most of them turn on as soon as the power failure occurs—no matter how brightly the sun might be shining at the time. When night falls and

Fig. 2-12. Two-function fluorescent light and flashlight (courtesy Radio Shack, division of Tandy Corporation).

the light is really needed, the battery may be all but dead, having served its purpose when no one needed it. The Emergency Light Source (ELS) you can build yourself overcomes this shortcoming. Shown in Fig. 2-13, the ELS has its own built-in electric eye (photocell) that senses ambient illumination levels. It does not allow the light to go on when there is sufficient ambient light for maneuvering about. Thus, the light will go on only when it is dark near the ELS. The photocell tells the box when to turn on.

The unit is designed with relay-controlled external terminals that are used to remotely indicate that a power failure has taken place even though the light does not come on. This important safety feature can be used for remote alarms, telephone self-dialers, or whatever purpose you choose. An alarm such as this is handy to have in a hothouse to warn of a power failure. In the event an intruder were to shut off your power switch, you would be alerted remotely that this had happened. The same holds true for deep freezers that might be located remotely from your house or work area. You would be alerted that power had failed to that area of your property.

How The ELS Works

The Emergency Light Source makes use of two 6-volt motorcycle storage batteries connected in series to provide 12 Vdc to power the light. The schematic diagram for the ELS is shown in Fig.

Fig. 2-13. Emergency Light Source with photocell for activation only when darkness prevails.

2-14. All parts will be available at your local electronics or automotive supply stores. The batteries are kept on a trickle charge as long as the power remains on and relay K1 is kept pulled in. The trickle charge voltage is provided by the secondary of transformer T1 and is rectified by diode D1. At the same time, a dc voltage is applied to relay K1 through diode D2 and resistor R3 to hold the relay "in." Thus, when the power is available from the 120-Vac line, the relay is held closed and the batteries are kept on trickle charge. In the event of a power failure (or when the plug is removed from the wall), the relay "drops out" and the batteries are now connected to the lamp I1 through the silicon controlled rectifier (SCR at Q1) and photocell combination light-sensing circuit. At the same time, the connections on the external terminal strip TB1 are switched on. If there is sufficient ambient light at this time, the lamp will not go on. If light is needed, the photocell triggers the lamp on through the SCR (Q1). Whenever there is no power at the 120-Vac input, the connections to the external circuit are made and a remote light, buzzer, or bell can be turned on or off by the three leads (single pole, double throw switch) of terminal board TB1.

The two 6-volt motorcycle batteries have a capacity of 2 ampere-hours. Of course, batteries of higher ampere-hour (AH) capacity can be used if higher output capacity is required. In Fig. 2-15, we see one of the batteries removed from the ELS and can also note how the various components are mounted to the case. Nickel cadmium (NiCad) batteries may also be used. The No. 67 lamp draws approximately 350 mA from the battery. Therefore, this battery-bulb combination should provide about six hours of useful light before the batteries need to be recharged. Most power failures in the United States do not last this long; many last from just a few minutes to perhaps several hours. To obtain more initial output with the same batteries, and with attendant shorter useful time, a higher current automotive lamp can be used. Of course, large batteries and a higher current lamp will provide longer hours of operation with higher light output. For the purpose for which this unit is intended, however, the selected battery-bulb combination should be sufficient. Once a power failure has occurred and the ELS turns on, the plug can be removed from the wall and the ELS used as a portable light source wherever needed.

Construction of the ELS

A medium size (4½ × 4 × 6½ inches) metal index card file cabinet houses all the components as shown in Fig. 2-16. The

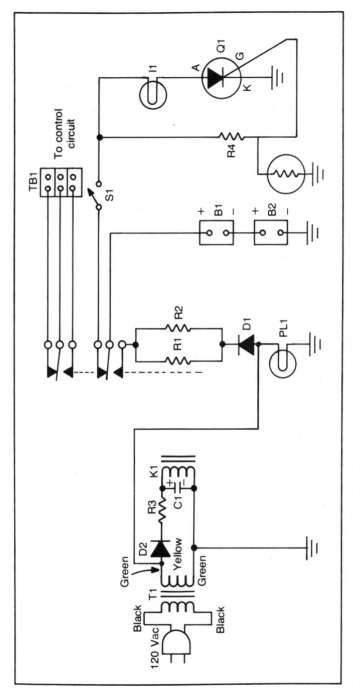

Fig. 2-14. Circuit diagram for Emergency Light Source with relay K1 deactivated.

31

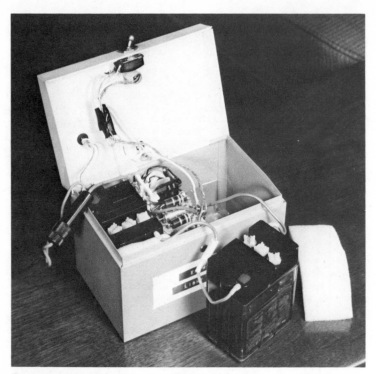

Fig. 2-15. Motorcycle battery removed from ELS unit.

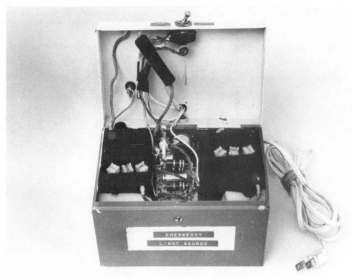

Fig. 2-16. Interior view of Emergency Light Source.

32

Table 2-3. Parts List for the Emergency Light Source You Build Yourself.

Item	Description
C1	Capacitor, 220 μF, 35 Vdc, electrolytic Radio Shack No. 272-1017, or equivalent
K1	Relay, DPDT, 12 Vdc, Radio Shack No. 275-206, or equivalent
R1, R2, R3 R4	Resistor, 33 ohms, 5 percent 2 watts
D1, D2	Silicon rectifier diode, 1 ampere, 1N4001, 40 PIV, Radio Shack No. 276-1101, or equivalent
Q1	Silicon Controlled Rectifier, Radio Shack No. 276-1067, or equivalent
T1	Filament Transformer, 12.6 Vac, 1 ampere, secondary. Radio Shack No. 273-1505, or equivalent
I1	Lamp, 12 Vdc GE 67, automotive marker lamp
PC1	Photocell, Cadmium Sulfide (CdS), Radio Shack No. 276-116, or equivalent
TB1	Barrier Strip, Radio Shack No. 274-657, or equivalent
S1	SPST Toggle Switch, Radio Shack No. 276-612, or equivalent
PL1	Pilot light with mounting, 12 volt, Radio Shack No. 272-334, or equivalent
B1, B2	Lead-Acid motorcycle storage battery, 6 Vdc, Yuasa, MBC 1-6
Case	Index Card File, 4½ × 4 × 6½ inches, or equivalent
Lamp-holder	Automotive marker lamp, Paterson 2c6306 or equivalent
Misc.	ac Line Cord, perfboards, nuts, screws, pill container

transformer is held to the bottom of the case with two screws. The diodes, resistors, and capacitors are held in place on a small perfboard that is supported above the transformer. The relay is mounted to the rear surface of the case with its own mounting screws. The batteries are held in place with foam rubber padding. Higher output batteries can be mounted in a larger box or mounted apart from the electronic circuitry. The clearance light assembly with its clear lens and switch (S1) to turn the light on and off are mounted on the cover of the hinged box. The pilot light on the cover indicates immediately that power is being supplied to the ELS and that the batteries are being kept in readiness.

The photo cell sensing unit is mounted in a pill container 1 inch in diameter and 2½ inches long. The photocell is mounted on the cover of the container. Resistor R4 and SCR Q1 are mounted on a small piece of perfboard. The leads that come from the bottom of the pill container are made long enough to leave about 2 inches clearance before they go into the index card case. This clearance allows orientation of the photocell so that it can be pointed toward the illumination whose level is being sensed, such as the nearby window or door through which natural light is being admitted. The parts list used to construct the ELS is shown in Table 2-3.

Home Security and Energy-Conscious Lighting

3

We all like to believe that the cities, towns, and hamlets in which we live are safe. But more and more homeowners, apartment dwellers, and campers are searching for better security in their environments. In this chapter, we will look at some of the security devices modern electronics has brought us. Much of the same technology which brings us better security also saves us money, so we will consider energy-saving devices as well.

SECURITY LIGHTING

Some security lighting devices, such as the photocell to turn on lights as night time approaches, have been around for a number of years—but in industry, not in the average home. Other devices, such as the heat sensor to turn on a light, are recent developments and are just now becoming readily available at local stores.

Night-Guard

The Night-Guard uses a photocell to sense the amount of light available and to turn on any lamp plugged into it. Figure 3-1 shows the light sensor up close, with a circular opening for the photocell, a light sensitive resistor that has several million ohms resistance in the dark and just a few hundred ohms resistance in bright light. The sensor is placed at some convenient place where light can strike it, such as in a window as shown in Fig. 3-2.

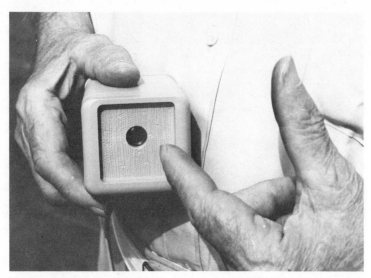

Fig. 3-1. Circular opening for Night-Guard's light sensitive resistor.

Fig. 3-2. Night-Guard placed in window exposed to outside light.

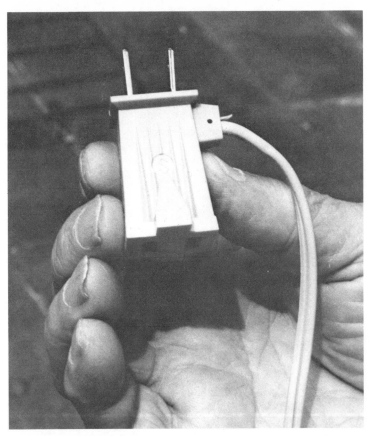

Fig. 3-3. Lamp to be turned on automatically is plugged into Night-Guard's receptacle plug.

The lamp to be turned on as dusk approaches is turned on and plugged into the Night-Guard's input to the wall outlet as shown in Fig. 3-3. When it gets dark, the sensor will turn the lamp on, to give your house or apartment a lived-in look. This will provide safety and security for your home when you are away.

The Night-Guard can be used with all table lamps and swag lights and is polarized for safety. The unit will switch up to 300-watt incandescent loads on and off. You can put one of the units in the living room and one in the bedroom. The sensor can be arranged to turn on and off at different times depending on the degree of darkness. You can place the Night-Guard on the floor for early turn on (it will be darker on the floor), on the window sill or table top for later turn on (it sees more light later), and you can face it toward or away from the window. Be sure that the sensor does not "see" the light from bulb it controls or else it will alternately turn on and off in the mistaken belief that the light from the bulb is daylight. Then when the lamp is off, it is dark again and the night guard once again turns on. The process repeats until daylight puts an end to this low-frequency oscillation. If you place a lamp close to the front door, you will eliminate the need to enter a dark home at night.

Night-Guard is manufactured under the Snapit logo by Cable Electric Products, Incorporated, P.O. Box 6767, Providence, Rhode Island 02940.

Lightwatch II

The Lightwatch II light control represents one of the latest advancements in passive infrared technology. Shown in Fig. 3-4, the Lightwatch II detects rapid changes in temperature (infrared energy) whenever a person, auto, or large animal moves through its invisible detection pattern. This pattern covers a distance of 40 feet and a width of 25 feet as shown in Fig. 3-5.

The Lightwatch II unit emits no radiation itself and is harmless to humans and animals. The sensor looks only for body heat. Each average-size adult radiates as much heat as that from a 100-watt light bulb. That is why a cool room filled with a crowd of people soon becomes warm and stuffy. A room filled with 100 people will have to dissipate about 10,000 watts!

This unit is designed to control up to a 500-watt noninductive lighting load (incandescent) in one or more light fixtures. The unit can control other devices but during normal operation, it will occa-

Fig. 3-4. Lightwatch II body heat detector and security lamps.

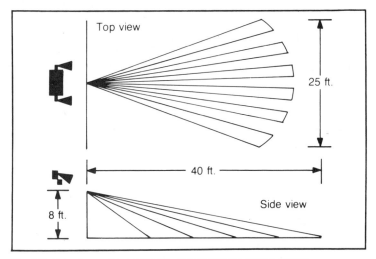

Fig. 3-5. Detection pattern of Lightwatch II infrared heat detector.

sionally activate due to rapid environmental changes, blowing bushes or movement of pets. Therefore, the unit should not be connected to any alarm system, siren, loud audio device, or automatic telephone calling device. The sensor is most sensitive to motion toward or away from the lens.

A Lightwatch II is shown in Fig. 3-6 installed under the eaves of a house looking out into the front yard. During daylight hours, a photocell transistor senses the ambient light and keeps the unit off. At night, the unit is activated automatically and anyone walking within the 40-foot beam will activate the unit. The incandescent

Fig. 3-6. Lightwatch II installed under eaves of a house. Body heat will turn lights on.

Table 3-1. Specifications for Lightwatch II.

Item	Description
Power requirement	120 Vac, 60 Hz
Power consumption	1-watt maximum (sense)
Detection range	40 feet × 25 feet
Detection method	Passive infrared (detects thermal radiation from heated objects)
Sensitivity	Adjustable over a 5-to-1 temperature differential range
Lights "on" time	Adjustable from 10 seconds to 20 minutes
Aiming capability	Fully adjustable
Switching capability	500 watts, incandescent

lamps will be turned on and then turned off automatically 10 seconds (or more) after the last person has left the area. The Lightwatch will automatically extend a warm greeting to your guests when they arrive at night and also welcome your own family when you come home after dark. Table 3-1 shows the specifications for the Lightwatch II.

The unit is manufactured by Colorado Electro-Optics, Incorporated, 2200 Central Ave, Boulder, Colorado 80301.

Light Alert

The Light Alert Outdoor Security System, similar to the foregoing unit, is a security device that turns on its own flood lights when it spots people or cars. When the Light Alert senses an intruder within its 40-foot zone of sight, it instantly blasts the area with blazing light. When lights suddenly come on, fully illuminating him and the local area, an intruder thinks he has been spotted and he makes an instant getaway. You do not even have to be awakened and yet you have been protected. It also avoids the problem of nuisance bells or sirens that can irritate neighbors. Remember—light is a silent siren and it blasts out a brilliant alert. Its wavelength is that of the eye, and not the ear. At other times, Light Alert provides a light welcome for family and friends.

When compared to all-night security lighting, Light Alert saves energy, and therefore, money, because it turns on the lights only when someone is there. The lights will stay on as long as there is body heat present in the watch area. The unit is adjustable to turn off automatically from 10 seconds to 20 minutes later. A built-in photocell in the Light Alert deactivates the unit during daytime hours. When activated, the unit will switch up to 500 watts incandescent.

The Light Alert is available as a complete ready-to-use package, including floodlights and sensor. The unit is manufactured by Light Force Security Lighting Systems, a division of RAB Electric Manufacturing Company, Incorporated, Bronx, New York 10451.

Passive Infrared Motion Sensor

A unit for use inside hallways or homes is the Passive Infrared Motion Sensor by Radio Shack shown in Fig. 3-7. This unit covers a 20-foot-wide-by-30-foot-long area and provides a loud internal audio alarm when the protected area is violated. It will alarm once and then reset. A 30-second exit/entry delay is provided. A battery backup system of eight C cells powers the unit during ac power failure.

The Passive Infrared Motion Sensor is carried as Catalog No. 49-305 by Radio Shack and is available at most local stores.

SafeHouse Passive Infrared Motion-sensing Head

This indoor wall-mounted motion-sensing head is made by Radio Shack and is shown in Fig. 3-8. It also senses body heat motion and covers an area of 35 feet long and 20 feet wide. It connects to a closed circuit loop and has a normally open (N.O.) switch which will sound an externally connected audio alarm or turn

Fig. 3-7. Passive Infrared Motion Sensor for use indoors (courtesy Radio Shack, division of Tandy Corporation).

Fig. 3-8. SafeHouse motion sensing infrared heat detector head (courtesy Radio Shack, division of Tandy Corporation).

on a light. A switchable walk-test LED indicator lets you determine the actual area of coverage. The unit requires 12 Vdc at 20 mA current.

The SafeHouse Passive Infrared Motion-Sensing Head is carried as Catalog No. 49-530 by Radio Shack and is available at most local stores.

Automatic Light Control

The Automatic Light Control shown in Fig. 3-9 is available from Radio Shack. With the unit you will be able to control a lamp so that it will turn on at dusk and off automatically when the sun comes up. You simply mount the light control in a window or on the table in an area reached by sunlight and plug a lamp into the unit. The unit's cord is then plugged into a 120-Vac outlet and it will control a maximum of 300 watts incandescent power.

The Automatic Light Control is carried under the Archer trade name by Radio Shack as Catalog No. 61-2774.

PHOTOCELLS, DIMMERS, DIODES, AND TIMING DEVICES

The practice of leaving lights burning when they are not needed can be eliminated, or power consumption minimized, by using photocells, dimmers, diodes, and timing devices. Most of these devices are solid-state and will operate for many years without replacement. In almost all cases, they will provide you with an energy savings and much less frequent lamp replacement. The vacationer who wants to give his home the appearance of being occupied, or the family that likes to find a light on when they return home will find these devices just right. And their cost is minimal compared to the money saved through energy conservation, convenience, and safety.

Photocells

The photocell offers an automatic way to turn lights on and off in direct response to the amount of natural light available. Photocells are small photo-electric or light sensitive resistors that are less than an inch in diameter and a quarter of an inch thick and are sensitive to sunlight or any bright light. Light striking a photocell is converted to electrical energy causing reduced electrical resistance across the device. The brighter the light the lower the electrical resistance.

Photocells are designed so that when a certain amount of light strikes them, the reduced electrical resistance caused by the light triggers an electrical circuit to turn off the lamp. When dusk ap-

Fig. 3-9. Automatic Light Control by Archer (courtesy Radio Shack, division of Tandy Corporation).

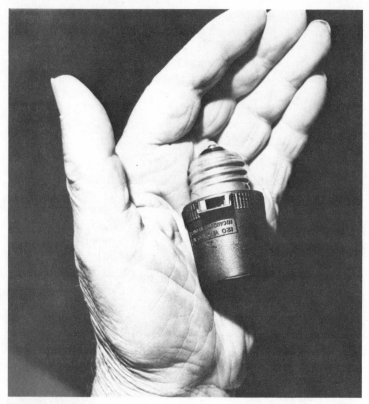

Fig. 3-10. Photocell light switch is screwed into base of lamp.

proaches, or even on dark stormy days when there is less ambient light, the photocell will switch on again. This turns the lamp back on.

Most of the devices are designed with a time-delay so that any sudden increase in light will not turn the lamp on immediately. Such might be the case of an outside front porch light when an auto's headlights momentarily illuminate the photocell. In the evening, the lamp may flicker on and off several times until the device knows that dusk and darkness have finally arrived. Then the lamp will stay on for the rest of the evening and night.

The photocell automatic lamp switch is shown in Fig. 3-10. This unit is screwed into the base of the lamp. The bulb is then screwed into the switch. There is a special lens opening which lets you rotate it in the direction of the outside light which will switch the unit off and on at dawn and dusk. In Fig. 3-11, we see the light-sensitive switch installed in the base of a lamp and the lens opening rotated in the direction of a bright window.

A light-sensitive switch available from Radio Shack as Catalog No. 61-2775 is shown in Fig. 3-12. In this photo, you can see the light sensitive resistor through the round opening on the side of the device. The opening can be rotated with the light and switch

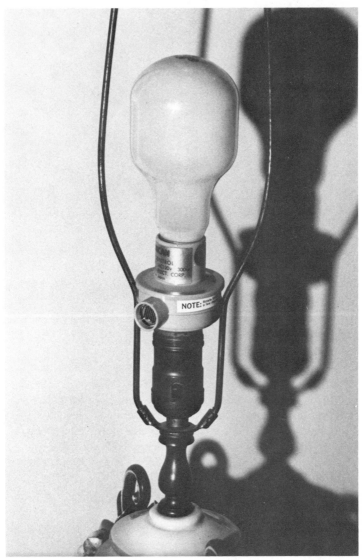

Fig. 3-11. Light switch installed in base of lamp. Lamp will come on automatically at dusk.

Fig. 3-12. Light switch lens opening can be rotated for optimum performance (courtesy Radio Shack, division of Tandy Corporation).

screwed into the base of the lamp. This unit can be used outdoors in any position and is ideal for driveway entrances, patios, or anywhere you need automatic light turn-on at night. The unit will fit floodlight sockets and will switch a maximum of 150 watts.

The light-sensitive switches described above will switch the light from fully on to fully off, with no in-between levels. This means a 75-watt light will be at its full brilliance when on, a 40-watt light at 40 watts, and so forth.

Dimmers

All incandescent light sources can be dimmed by controlling the amount of voltage applied to them. A dimmer control permits a person to control and adjust the level of light in a room or work area over the full range, from off to the highest illumination level. There is no doubt that dimmers can achieve an energy savings when they are used regularly.

There are several types of lamp dimmers on the market. Most of these replace standard wall light switches and are inexpensive to buy and easy to install.

Rotary Dimmer. The wall-mounted light dimmer shown in Fig. 3-13 is a push on/off rotary dial which varies the light level to

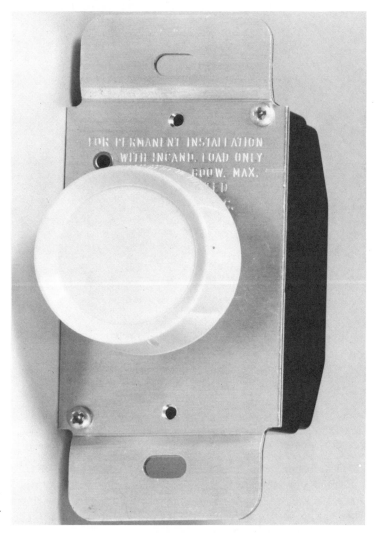

Fig. 3-13. Rotary lamp dimmer with push in-out for on-off control of room lights (courtesy Radio Shack, division of Tandy Corporation).

Fig. 3-14. Toggle dimmer wall switch provides full-off and full-on with full-range dimming in between (courtesy Radio Shack, division of Tandy Corporation).

suit the occasion. It fits the standard wall box and controls up to 600 watts of incandescent lights at 120 Vac. These dimmers should not be used for controlling the speed of motors or such devices such as saber saws. They can be used to control the temperature of a soldering iron as it is a resistive device.

Toggle Dimmer. The toggle dimmer shown in Fig. 3-14

serves the same function as the rotary dimmer except that the familiar toggle is used to provide down-off, up-on control of the brilliance of the light, plus full-range dimming in between the two positions. The rating for this switch is also 600 watts incandescent load and is also installed in the familiar wall switch box.

Fluorescent Lamp Dimmers. Dimmers for use with fluorescent lamps must be used with a special dimmer ballast, which replaces the standard ballast. Since the use of fluorescent lamps in itself provides energy savings, the use of dimmers with fluorescent lamps is not as widespread. Dimming equipment is available for only 30-watt and 40-watt rapid-start fluorescent lamps. Even when used, there is a tendency for the fluorescent lamps to flicker at the lower light level settings and this may prove to be distracting.

Diodes

These tiny, solid-state devices are a good way to save energy and lamp replacement costs. A diode is a solid-state rectifier that can be placed between the base of a lamp and the bulb. The diode converts alternating current voltage into pulsating direct current voltage. Incandescent bulbs can last up to 100 times longer when used with a diode. This means that a 75-watt bulb with an average 750-hour lifetime can burn for about 75,000 hours, or 24 hours a day for eight years!

Because the bulb burns cooler with a diode, the amount of light it puts out is slightly reduced and the color appears slightly yellow. To get more light, you can use a larger size bulb and still have sufficient energy because color rendition means little when employed in exit signs or walkway and driveway background lighting.

Diodes and lamp dimmers extend the life of the incandescent bulb because they reduce the amount of inrush current that flows into a cold filament when the bulb is first turned on. Inrush current is normally 12 to 15 times as great as the operating current. This is shown in Fig. 3-15. Diodes also extend the life of the filament by operating it at reduced voltage over a good portion of its lifetime. This can extend a 10-week bulb lifetime into a 10-year lifetime! The curves in Fig. 3-16 will give you an idea of how the lifetime increases when, light output decreases with a decrease of voltage. Note that at 70 percent of rated voltage, light output in candle power is reduced to 25 percent, but average lifetime is increased over 100 times! This means that a 100-watt bulb might look only as bright as a

25-watt bulb but it would last for 75,000 hours instead of just 750 hours!

Diode bulb savers may be ordered from Electronic Supermarket, P.O. Box 988, Lynnfield, Massachusetts 01940, using Catalog No. 3VL0231, or from Brookstone, 127 Vose Farm Road, Peterborough, New Hampshire 03458, using Catalog No. H-9181.

Sensor Timer

The Sensor Timer is a novel solid-state timer compared to mechanical motor timers. The unit has been designed to provide the safety and security of automatic lighting without the inconvenience of mechanical timers. The Sensor Timer combines the features of several devices discussed earlier—turning lights on and off so that a residence appears to be occupied, and to do this only at night when the feature is required.

The Sensor Timer uses a photoelectric eye to sense the presence or absence of natural daylight to turn the lamp on so that you never need to make seasonal adjustments for daylight and time of day as you would a mechanical motor-operated clock timer. Because the unit does not depend on the time of day for proper operation, power failures do not upset the timing cycle as with an

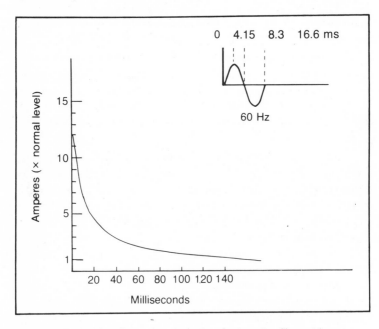

Fig. 3-15. Typical inrush current versus time for tungsten filament lamp.

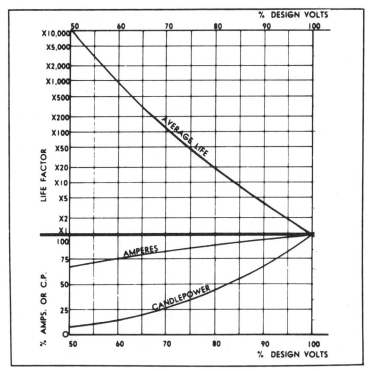

Fig. 3-16. Average life versus voltage applied to incandescent lamps.

electric clock motor timer. The Sensor Timer returns to automatic operation after a power failure.

Two operating modes are provided with the Sensor Timer—fixed or random. In the fixed mode, the timer will turn the light on for a fixed period of time (selectable with a slide adjustment) for 2, 4, 6, 8 or 10 hours. When set in random, the solid-state timer will turn the light on for periods of 20 to 60 minutes and off for periods of 10 to 20 minutes. This will give that room of your house a very lived-in appearance.

The Sensor Timer is manufactured by Diablo Technologies, Incorporated, 5115 Port Chicago Highway, Concord, California 94520.

THE INTRUDAFLASH YOU CAN BUILD YOURSELF

The Intrudaflash is a simple, reliable, and economical intrusion alarm that can give you protection against intruders. It is designed to be activated by passive switch sensors at each door and window, or by manually actuated "panic" button switches placed strategically through your home. Once activated, the Intrudaflash will turn on a bright light and then off in rapid sequence and sound a loud bell to summon aid to your home. This activity will help put your intruder to flight.

How It Works

The Intrudaflash goes into action the instant contact is established across the *sense* terminals which connect to the control winding of the ac-line, isolation relay K1, Fig. 3-17. It begins to apply pulsed 115 Vac to any combination of lights, bells, sirens, horns, or buzzers, that are plugged into the *load* outlet mounted on the side of the plastic utility case which houses the Intrudaflash.

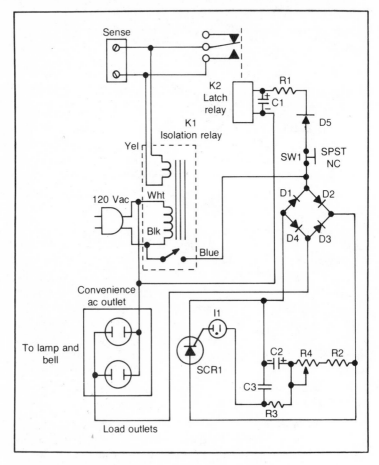

Fig. 3-17. Schematic diagram for building the Intrudaflash.

At the same time that K1 applies 115 Vac to that portion of the Intrudaflash circuit which produces the flashing action, it also applies voltage to dc relay K2 so it also closes. Diode D5 rectifies and resistor R1 limits the current through this relay to a safe value. Capacitor C1 provides the filtering necessary to avoid relay chatter of K2. The normally open contacts of relay K2 are connected across the control winding of relay K1 so they are effectively in parallel with the sense terminals. This arrangement latches the unit On and guarantees that the flasher circuit continues to operate once it has started, even if the connection between the sense terminals is opened.

The four rectifier diodes D1, D2, D3, and D4 form a bridge rectifier circuit and the SCR is connected across the dc legs of this bridge. As soon as ac is applied to this bridge, capacitor C2 starts to charge through the series combination of fixed resistor R2 and variable resistor R4. Capacitor C3 now also begins to charge through R3. When the voltage across C3 reaches about 70 Vdc, it fires the neon bulb which then discharges C3 into the gate of the SCR. The SCR conducts and applies full line voltage of 115 Vac to any load plugged into the Load outlet. As soon as the SCR conducts, C2 starts to discharge through R2, R4, and the SCR. This action provides a dc current through the SCR which is above the holding current, so the SCR stays on. When the current available from C2 drops below the holding current of the SCR, it will turn off during the next ac cycle when the line voltage is near zero. With the SCR gate "open," C2 starts to charge once again through R2 and R4 until the neon bulb fires once again and the cycle is repeated. The 5,000-ohm potentiometer, R4, provides for a variable flash rate from about 15 flashes per second at its maximum resistance setting to 80 flashes per second at its minimum resistance.

To stop the flashing, simply press the pushbutton switch, SW1, a normally closed SPST, which interrupts the circuit to K2 so it opens. This removes the connection across the control winding of K1, and unless there is still a connection across the sense terminals, the action stops.

Construction

To provide for safety while in use, the Intrudaflash is housed in a bakelite utility case that is closed with a bakelite cover. All of the component parts, with the exception of the isolation relay, K1, are mounted on one side of a 3¼ × 4 inch perfboard which is held in place by two machine screws. Isolation relay K1 is mounted next to the perfboard and is held in place by two 1¾ inch 6-32 screws. The sense terminal strip, which provides the "input" is located on one side of the case. The utility outlet, which furnishes the output, is affixed to the opposite side of the case. The rate adjustment potentiometer, R4, and the reset pushbutton, SW1, are mounted on the bakelite cover. Be careful to locate these two components in such a way that when the cover is screwed in place neither the potentiometer nor the pushbutton short against any of the components on the perfboard or against one of the relays.

Sensors and Wiring

Switch-type sensors should be placed at each door or window you wish to protect. You can use magnetically activated reed switches, or even "homebrew" switches consisting of a wiping contact mounted on the door and a fixed contact on the door jab. The switch you choose should be open (no contact) when the door or window is in its usual, closed position. When the window or door is opened, the switch contact is made (closed) and the alarm starts.

For added detection capability, you may want to consider use of pressure-sensitive mats or strips under rugs and carpeting. These are especially good choices for in-house protection, just in case someone should get past the perimeter protection on doors and windows.

Of course, you can also add photoelectric or ultrasonic sensors to the system as long as the unit supplies a contact closure when it detects an intruder. Most sensor units give you the choice of normally open (N. O.) or normally closed (N.C.) contact outputs.

Do not forget to provide one or more "panic" switches at strategic points around the house. Consider locations such as next to the bed, in closets next to the front and rear doors, and in the kitchen or laundry room. The objective is to place means of activating the alarm in locations that can be reached easily and quickly, just in case an intruder attempted to enter your home during daylight hours when the family is there but the alarm is essentially "off."

Wiring to the sensors and panic switches consists of two leads which carry a very low voltage, safely isolated from the power line by the transformer action of the isolation relay K1. All sensors and switches are simply wired across the two leads so that a momentary closure of any pair of sensor contacts will shunt the two wires with a very low resistance, which sets the alarm off. Ordinary colorless flat-ribbon TV lead-in can be used for wiring if you wish. It is very easy to pass under carpeting, through walls, or along baseboards

Table 3-2. Parts List for the Intrudaflash You Build Yourself.

Item	Description
C1	25 μF, 100 Vdc electrolytic capacitor
C2	100 μF, 150 Vdc electrolytic capacitor
C3	0.1 μF, 200 Vdc Mylar capacitor
CS1	Case, plastic, 6 × 3 3/16 × 1 7/8 inches Radio Shack No. 270-223, or equivalent
D1, D2, D3 D4	10 amperes, 200 PIV rectifier diode
D5	1 ampere, 200 PIV diode
I1	Type NE 2H neon bulb
K1	Alcoa Model FR-105 SPST isolation relay
K2	Mini relay, Radio Shack No. 275-004, 500 ohm coil
R1	3900 ohm, 10 percent 5-watt resistor
R2	500 ohm, 10 percent 5-watt resistor
R3	1 megohm, 10 percent ½-watt resistor
R4	5,000-ohm potentiometer
SCR1	Type C30B silicon-controlled rectifier
SW1	SPST normally-closed pushbutton
Misc.	Line cord, duplex utility outlet, terminal strip, 3½ × 4 inch perfboard, knob

since it can be tacked or stapled to surfaces very easily. Bell or speaker wire are also good as long as the wire you choose has a low resistance, and is at least #18 AWG or larger in size (smaller #). Small-diameter wire may lead to unreliable detection at distances of several hundred feet because of its high series resistance. Remember, thin, small wire is fragile and easily broken. The parts list for the Intrudaflash is shown in Table 3-2.

Security Lighting and Safety Away From Home

4

When traveling by auto, camper, recreational vehicle, bike, or just plain hiking, we feel better if we know that we can adequately cope with an emergency should it arise. We feel better when we have a good spare tire, a full tank of gas, and a good plan to arrive safely at our destination. This chapter will cover several aspects of traveling, hiking, or camping safety as relates to lighting and the assurance we can have if we have some of the devices to be described.

COMMERCIALLY AVAILABLE DEVICES

The next four subsections describe products useful in a variety of situations from car breakdowns to camping. Specifics about product size and application will help you make an informed selection when you want to buy one of these safety devices.

Auto-Alert Flashing Fluorescent Lantern

This dual-purpose light by Nicholl Brothers, the F-1F, is indispensable for scores of routine and emergency uses. Figure 4-1 shows the F-1F four-in-one fluorescent light which can be used as a flashing distress signal, as a floodlight for home, shop, or auto, as an anti-bug light for patio and camping with its amber filter in place, and as an amber hazard warning flasher. The F-1F uses a heavy-duty 6-volt lantern battery and a 6-watt fluorescent bulb to provide a bright floodlight for lighting the camp area or for flashing emergency messages to passing motorists should your vehicle break down.

Fig. 4-1. F-1F four-in-one fluorescent light.

Figure 4-2 shows a "HELP" message placed across the lantern face and three other messages which could also be used. The F-1F message system operates in the following manner. Select the message you need from the handy assortment packed with the lantern or use the amber filter also packed with the F-1F to signal a flashing caution. Turn on the lantern and start it flashing. You can lay the lantern on its side for easy readability of the message and put it wherever it will be most visible. Then when help arrives, you turn off both switches and return the message to its package.

The F-1F's multi-function handle allows the lantern to be carried in two comfortable positions. It can also be hung from a tent or tree branch in any of four positions you choose. A special slot lets you hang the F-1F on a wall, such as in your garage, from a single nail or screw.

Specifications for the F-1F are shown in Table 4-1. The Flashing Fluorescent Lantern is made by Nicholl Brothers, Incorporated, 1204 West 27th Street, Kansas City, Missouri 64108.

Fig. 4-2. HELP message placed across face of F-1F.

Personal Protection Light

Made by ATGM Enterprises, Incorporated, the Personal Protection Light provides for your safety by blinking 80 times per minute. The unit is lightweight, weighing under 4 ounces with the batteries. It is compact, and clips to your clothing so you can use it when jogging, walking, camping, boating, biking, repairing your auto on the road, or even when working as a school-crossing guard.

Table 4-1. Specifications for F-1F Flashing Fluorescent.

Item	Description
Case	Rugged high-impact injection molded
Circuitry	100 percent solid-state
Lens	Unbreakable polymer
Bulb	F6TSCW 7,500-hr. rated, F6T5BLB Black Light available
Reflector	Super bright metallic finished Mylar
Battery	Uses one standard or heavy-duty 6-volt lantern battery
Battery life	Intermittent 14-17 hrs. or 8-10 hrs. with heavy duty
Switch	Trigger type rocker protected against bumping and moisture
Handle	Wrap over type with screw, nail slot
Size	9½ inches high × 3⅛ inches wide × 6½ inches long
Weight (w/o signs)	1 pound ¾ ounces, 2 pound 6 ounces with battery

Fig. 4-3. Personal Protection Light by ATGM Enterprises, Incorporated.

The personal light can also be switched to continuous lighting with the operating switch, which has three positions: off, blinking, and continuous.

The ATGM light is shown in Fig. 4-3. The large lens surface is 3⅝ inches in diameter and is made of tough weather-resistant plastic. The light from the lens is bright enough to be seen up to a mile away. The protection light uses two AA alkaline cells and will blink for more than 20 hours. The solid-state integrated circuit and batteries are shown in Fig. 4-4, where the reflective lens has been removed by clasping the unit between both hands and twisting the lens off. The bulb is easily replaced once the lens is removed.

The Personal Protection Light is available from ATGM Enterprises, Incorporated, 2810 Morris Avenue, Union, New Jersey 07083.

Polaroid Safety Flasher

Polaroid Corporation makes a lightweight hazard or emergency warning device which emits an orange blinking light visible at night for more than one mile. This light, called the Polaroid Safety Flasher, is a compact unit capable of flashing for up to four hours on current provided by a single Polaroid Polapulse P100 6-volt planar battery. The flasher is shown in Fig. 4-5 sitting on a flat surface. The safety flasher provides a simple, inexpensive, and convenient way for motorists to warn approaching vehicles of an accident or breakdown, or for boaters, campers, and cyclists to signal for assistance. Parents who want to let their children know they should come home can place the flasher in a window of the house. It is a safe substitute for flares since it can be used in many ways that flares cannot.

A short clockwise turn of the safety flasher's orange beacon lens switches on the unit to emit a high-intensity 1/10 second flash at approximately 2-second intervals. The safety flasher's tough, reflective, weather-resistant casing incorporates a collapsible

Fig. 4-4. Two AA cells power ATGM flasher light.

Fig. 4-5. Polaroid Safety Flasher with orange beacon lens.

metal handle which allows it to be hung, carried by hand, or strapped to the rear of a bicycle. It can also be set up quickly on a flat surface such as a tabletop or on the top of an auto. Suitable for convenient storage under the car seat, in a first-aid kit, or knapsack, the unit weighs less than 4 ounces.

Designed to accommodate Polaroid's wafer-thin Polapulse P100 battery, shown in Fig. 4-6, the compact Safety Flasher mea-

Fig. 4-6. Polapulse P100 6-volt battery pack.

Fig. 4-7. Polapulse battery pack being inserted in Safety Flasher.

sures only 4¼ × 3 7/16 inches and is 1⅛ inches thick. The Polapulse battery can be replaced in seconds as shown in Fig. 4-7 where a new battery is being inserted into a slot at the base of the unit. A snap-fastener provides quick access to the lamp holder for bulb replacement as shown in Fig. 4-8 where the orange lens cover has been removed from the unit.

The manufacturer recommends that you not store the safety flasher or P100 battery in automobile trunk, glove compartment, boat engine compartment, or surface areas which are subject to excessive heat. The convenient hook on the storage pack can be used to store the unit securely under the automobile front passenger seat as this is one of the coolest storage locations in an automobile.

Safety Flashers and Polapulse batteries are sold at automotive parts stores, and other retail outlets.

Security Strobe

Figure 4-9 shows a Security Strobe light made by Radio Shack which puts out a brilliant yellow burst of light. The unit uses a strobe light which operates off 12 Vdc and flashes at the rate of about once per second. The strobe is visible even in fog and indicates where an alarm is coming from and helps police or fire department locate your home quickly. The unit is water-proof for outdoor use and has a peak flash of 100,000 candlepower. The strobe light has

Fig. 4-8. Safety Flasher lens cover removed showing bulb replacement.

Fig. 4-9. Security Strobe beacon operates off 12 Vdc and has 100,000 candlepower flash (courtesy Radio Shack, division of Tandy Corporation).

screw terminals on the bottom of the unit for power application as well as two mounting holes to secure it to any desired location in your vehicle or residence. The strobe unit draws 180 mA at 12 Vdc and is UL-listed.

Radio Shack carries the strobe flasher as Catalog No. 49-527.

DO-IT-YOURSELF PROJECTS

You can complement your stock of helpful commercial products with safety devices you can assemble on your own. The following project descriptions offer tips on building and using some simple but effective devices.

Auto Trunk Flasher Light

A flasher light which mounts on your automobile trunk lid and folds inside when you close the lid is shown in Fig. 4-10. This red flasher light supplements the 4-way flasher lights on your car, and with the trunk lid open, stands almost 6 feet in the air where it can easily be seen. The light is built from a high-intensity lamp which you use at home and would operate off 120 Vac. What we will do is to remove the transformer from the high-intensity lamp, replace it

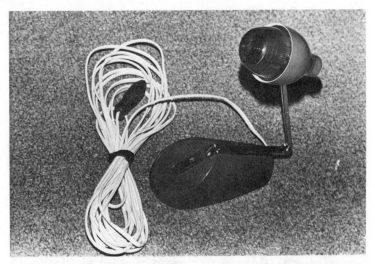

Fig. 4-10. Auto trunk flashing light you can build yourself using high-intensity unit.

with a three-prong directional flasher, and then power it off the 12 Vdc from your car's cigarette lighter socket.

The schematic diagram for the flasher light is shown in Fig. 4-11 and is very simple to wire up. A bottom view of the base of the

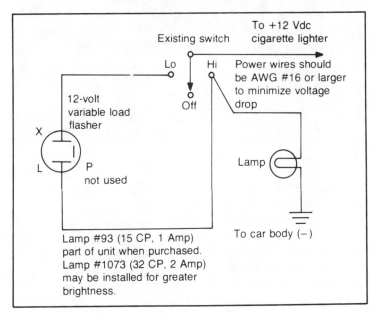

Fig. 4-11. Schematic diagram for auto trunk flasher light.

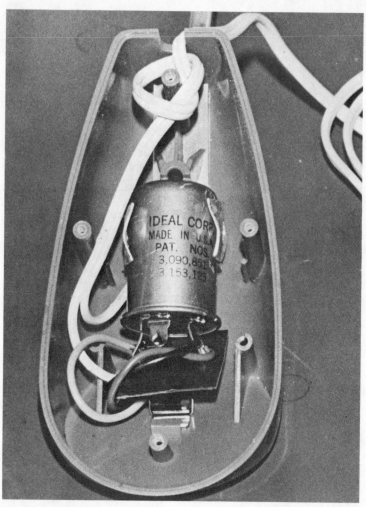

Fig. 4-12. Bottom view of trunk flasher light.

flasher lamp is shown in Fig. 4-12, along with the power cord that is plugged into the lighter socket.

To build the flasher lamp, follow the schematic diagram. Clip off the wire leads close to the transformer so that you can use them again. In place of the transformer, mount a three-prong 12-volt four-way directional flasher available from any local auto parts store. You will use the same high-intensity lamp, which is a No. 93, or a No. 1073 bulb, which will give you twice as much light. In addition, you will need a 2½-inch diameter red plastic truck clear-

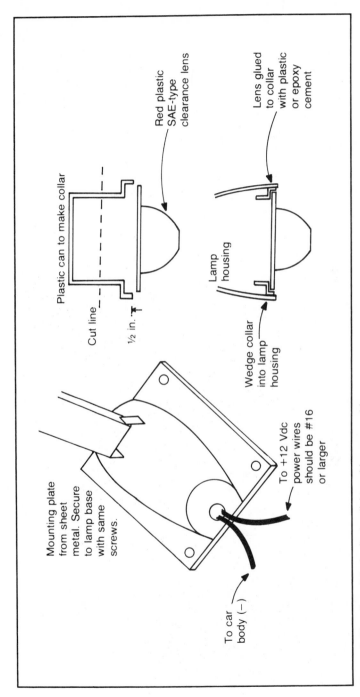

Fig. 4-13. Mounting plate for flasher light and lens assembly.

Red plastic
SAE-type
clearance lens

Lens glued
to collar
with plastic
or epoxy
cement

Plastic can to make collar

Cut line

½ in.

Lamp
housing

Wedge collar
into lamp
housing

Mounting plate
from sheet
metal. Secure
to lamp base
with same
screws.

To +12 Vdc
power wires
should be #16
or larger

To car
body (−)

ance light lens, SAE No. 57. A 6-inch square mounting plate cut from sheet metal is used to secure the lamp inside the auto trunk lid cross members as shown in Fig. 4-13. Details on securing the red lens to the lamp housing are also shown. The lens and bulb are seen in Fig. 4-14.

The original 3-way high-intensity light switch positions of Off-Low-High now become Off-Flash-Continuous. In operation on the highway roadside, open the trunk lid, move the flasher light into position, uncoil the power wire and plug it into your cigarette lighter socket. You will now have additional safety while you make necessary repairs. The flashing red clearance light will enhance the visibility of your auto as seen from a distance by oncoming motorists.

Headlight Modulator for Motorcycles

In order to increase motorcycle rider safety, the National Highway Traffic Safety Administration has proposed a safety standard amendment that would allow installation of a modulating headlamp on motorcycles. Just as railroad engines swing their headlamps back and forth, and in a circle, across the railroad tracks to warn of the approaching train, the motorcycle would switch its beam from low to high intensity at a certain rate visible to the human eye. Such a headlamp could considerably improve the visibility of the motorcycle during daylight hours.

Fig. 4-14. Lens and bulb for trunk flasher light.

The proposal would allow a normally steady burning headlamp to be wired to modulate the upper or bright beam from a high intensity to a low intensity at a pulse rate from 150 to 240 times per minute. At a rate of 240 flashes per minute, one would see a flash rate of four per second as a motorcycle approaches with its headlights on.

Circuit Operation. A circuit that will switch the high beam on and off at a four-per-second rate is shown in Fig. 4-15. The circuit is based on a 555 integrated circuit (IC) chip timer operating as a slow astable square-wave generator. The ratio of R1 to R2 is selected to produce a nearly symmetrical (squarewave) waveform. Timing capacitors C1 and C2 charge through R1 and R2 and discharge through R2 only. The output, taken from pin 3 of the 555 IC chip, is nearly square and the positive-going portion of the pulse turns on switching transistor Q1. Q1 in turn supplies the high current to the high beam filament of the headlamp.

Diode D1 passes only the positive-going portion of the squarewave to Q1. When Q1 is turned on, it acts as a switch and

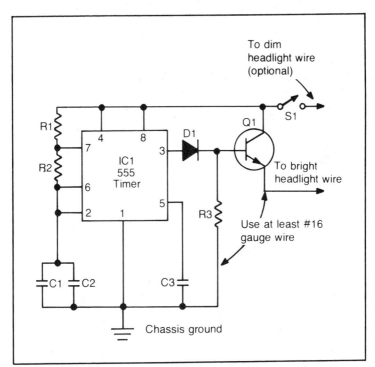

Fig. 4-15. Circuit diagram to build a motorcycle headlamp modulator.

Table 4-2. Parts List for Headlight Modulator You Build Yourself.

Item	Description
C1, C2	Capacitor, 1-μF 35 volt tantalum
C3	Capacitor, 0.01-μF, 50 volt disc ceramic
D1	1N914 Diode
IC1	555 timer IC
Q1	Transistor, ECG 182 (50-watt headlamp) or ECG 184 (30-watt headlamp)
R1	Resistor, 1 kilohm, ½ watt
R2	Resistor, 100 kilohms, ½ watt
R3	Resistor, 100 kilohms, ½ watt
S1	Switch, SPST (optional), 5-A contacts
Misc.	Perfboard 1½ × 1½ inches, wire, solder, 8-pin IC socket, 16 gauge wire, mounting hardware

provides the 12 Vdc to the headlamp. When Q1 is switched off, only the low beam filament is on.

Power to the modulator circuit comes from the low beam circuit. When connected as shown, the high beam indicator on the instrument panel will flash in unison with the high beam and will serve as a visual indicator that the unit is working. Switch S1 is optional and is used to turn the modulator off when it is not desired.

Construction. The unit can be mounted in a small plastic box or coated with a good silicon sealer and installed inside the headlamp housing if its size is kept small enough. Choose almost any npn power transistor for Q1 as long as it can handle the required lamp current of 5 to 7 amperes for a moped or small bike and about 10 amperes for a 50-watt headlight. Use a heatsink for Q1 to help dissipate the heat generated. All connections to the headlamp can be made without exposing any wires by placing all components within the headlamp housing. If optional switch S1 is used, it can be mounted on or near the headlamp.

Operation. When the headlights are turned on and dimmed, the modulator starts operating and the high beam is switched at a four pulse-per-second rate. Most motorcycles are equipped with the headlight switch having a center position when both high and low beams are on at the same time. When this center position is used, the modulator is switched off.

A parts lists for the headlight modulator is shown in Table 4-2. All components should be available at your local electronics store.

Light-Sensitive Security Alert

This is a very simple, unique device that uses a modern-day solid-state electronics to provide you with security information surrounding your home, camp, or recreation area premises. The ever-increasing interest in security and security devices has led to the development of a number of different systems. Most of these have been based on ultrasonics, interrupted light beams (visible or IR), normally open and normally closed switches, and conductive tapes placed on windows and doors. All except the simplest use one or more ICs as the alarm generator and for timing in various phases of operation.

Recently, however, a unique type of IC motion detector was developed for use in security systems. This device is based on large scale integration (LSI) technology and includes both linear and digital circuitry in its operation.

This unusual motion detector operates on a change in light level. This technique is not to be confused with systems that require a light transmitter and receiver and operate only when the light beam is completely interrupted to produce a 100 percent change in the level of light falling on the receiver. The light sensor used here can detect a change in light level as little as 5 percent and does not require a separate light source!

The advantage of this system is that triggering of the unit results when any motion of a person or object such as a door in the immediate vicinity causes either the direct or reflected light to change as little as 5 percent. This change may be in either direction—more light or less light. The system operates in light levels varying over a range of 1000 to 1; from as low as 0.1 candlepower, quite dark, to 100 candlepower, quite bright. The light sensor is contained in a molded clear dual in-line plastic (DIP) housing so that light can reach an internal photodiode detector. A molded lens is mounted over the package to improve the sensitivity at low light levels. The lens and hole (aperature) in the side of a mounting case are positioned so the photodiode gathers light in the shape of a cone. The plastic lens gathers light in such a manner that the light covers a 2-foot circle at a distance of around 8 feet.

Figure 4-16 is the block diagram of the D1072 light sensor IC chip. Light falling on photodiode D1 causes a voltage to be developed at its output. Now, any motion within the diode's field of view that changes the light level produces a change in the output voltage. Capacitor C1, connected between terminals 6 and 7, couples the voltage changes to amplifiers A2 and A3. Capacitors C1

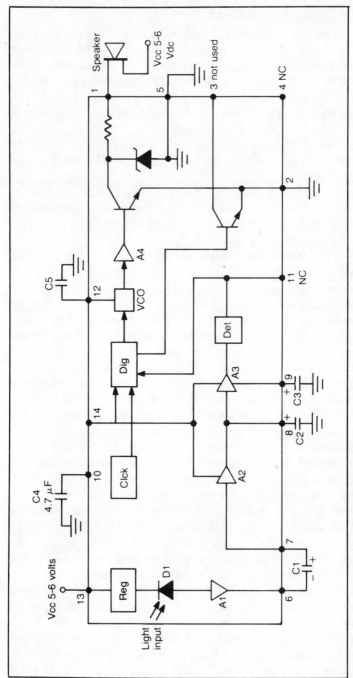

Fig. 4-16. Block diagram of the D1072 motion detector IC chip.

72

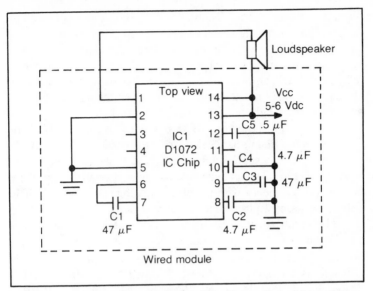

Fig. 4-17. Motion-detector module is complete with capacitors. Add power supply and small loudspeaker to complete operation.

and C3 are parts of a filter network to detect low-frequency voltage changes (motion) while C2 attenuates (filters out) higher frequencies, thus making the device immune to flicker inherent in lamps operating from ac sources (120 Hz flicker).

The gain of amplifiers A2 and A3, along with the log characteristic of A1, provides a signal that will trigger the internal detector when the change in light exceeds plus/minus 5 percent. The trigger produces a short tone burst from the internal generator. Capacitor C5 determines the frequency of the tones generated by the voltage-controlled oscillator (VCO) while capacitor C4 determines the ramp rate (rate of change) of the VCO's tone.

The audio alarm, once triggered, will continue for around 4 to 12 seconds, depending on circuit components. The sound is best described as a "whoop-whoop" sound. At the end of the alarm period, when the "whoop" stops, the light level is again sampled. If the light level is still changing, the audible alarm will continue. If the light level has stabilized—not necessarily at its previous level, which could have gone up or down—the alarm stops "whooping" and the circuit resets to detect the next change in light level.

Because the D1072 chip operates from a *change* in light intensity, it cannot operate reliably in very dark applications. For dark applications, use an invisible infrared (IR) source such as an LED or

73

filtered fluorescent lamp. The unit will operate extremely well in the daytime, looking for shadows that are cast any place in the room and which change the light level. It will signal any approaching cars in the driveway at night because the headlights cause an increase in light level. It will also signal a decrease in light level if an auto does not have its lights on but does obstruct some distant light source such as a street light or your neighbors yard light.

Figure 4-17 shows the IC chip and the capacitors that are mounted on the module as received from the vendor. All you have to provide is a 6-Vdc source and a small loudspeaker and the unit is ready to operate. The unit can be mounted in a small plastic utility box. Some experience with your local lighting conditions will help determine its best placement to suit your area.

The Light Sensitive Security Alert IC module is available as Catalog No. 7183 from Poly-Paks, Incorporated, 16-18 Del Carmine Street, Wakefield, Massachusetts 01880. The same module is available as Catalog No. D3AS0160 from BNF Enterprises, 119 Foster Street, P.O. Box 3357, Peabody, Massachusetts 01960.

5 Flasher Lights and Flashlights

Since the days of the cave dwellers, man has used all sorts of devices to help light his way at night. Over the eons, man has used burning bushes, bon fires, candles, kerosene lamps, whale oil lanterns, oil rich nuts strung together, oil-soaked sticks, twisted strips of birch bark, burning lumps of fat, palm tree torches, and even gourds full of holes filled with fireflies. Later on, lanterns that operated off gas served as portable lights. All of these means of lighting had their problems with rain, wind, snow, and sleet. The modern flashlight was slow in coming. First, the battery had to be invented by Alessandro Volta (1745-1827) and then the incandescent bulb had to be invented by Thomas Edison in 1879. But finally, the first dry-cell flashlight was made about 1898 in New York City. This chapter covers lights that flash and flashlights, and there are a variety of them!

BRINKMANN Q-BEAM SPOTLIGHTS

Brinkmann Corporation manufactures four spotlights for convenient illumination where you need it. Each light has a special feature, making it useful for a specific application.

Q-Beam Big Max

The Q-Beam Big Max is perhaps the world's most powerful portable spotlight that can be powered from any conventional 12-volt power source. Figure 5-1 shows the Big Max with an 8-foot

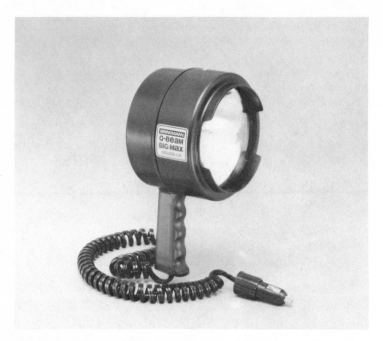

Fig. 5-1. Q-Beam Big Max has a 300,000-candlepower beam (courtesy Brinkmann Corporation).

coiled cord (Model 800-1901-0) which plugs into a cigarette lighter socket. Big Max packs the brilliance of 300,000 candlepower with safety and convenience. It will illuminate objects up to two miles away and is eight times more powerful than an average auto headlight on high beam. It projects light 500 times brighter than a two-cell flashlight.

The Q-Beam Big Max will operate off any 12-volt automotive, marine, agricultural, or industrial power source. It connects to the cigarette lighter socket or hooks directly to a 12-volt battery with the Q-Beam optional pigtail adapter accessory. Figure 5-2 shows the Big Max being powered from a battery pack.

The Big Max will operate with no battery drain when the engine or generator is running. It is encased in super tough ABS scratch-proof and shock-resistant plastic, and will withstand up to 5 Gs impact without breakage. Big Max will float, work under water, and weighs only 2½ pounds. There is a protective lens guard and the heavy-duty switch is recessed to prevent accidental activation. The handle grip is designed for slipfree use. A stainless steel retractable hanger enables the light to be stored conveniently. The Big Max is available with a 15-foot straight cord as Model No. 800-1900-0.

Q-Beam Quartz Halogen

This quartz halogen sealed beam hand-held spotlight shown in Fig. 5-3 combines tremendous candlepower with very low amperage draw. With a current of only 3 amperes, it will produce 160,000 candlepower of light. It is recommended for use with the Q-Beam Power Pack. This quartz halogen light comes with an 8-foot coiled

Fig. 5-2. Big Max powered from 12-volt battery pack.

Fig. 5-3. Q-Beam Quartz halogen spotlight has a 160,000-candlepower beam (courtesy Brinkmann Corporation).

cord and lighter plug. It is Model No. 800-1551-0 and weighs 2½ pounds.

Q-Beam Sportlite

This yellow spot and flood light is shown with a power pack in Fig. 5-4. It has an 8-foot coiled cord with lighter plug. The Sportlite is Model No. 800-1303-0.

Q-Beam Magnetic Base Spot/Flood

For those that need a vehicle-mounted spotlight, the Magnetic Base Spot/Flood is just the thing. Figure 5-5 shows this unit mounted on top of a vehicle. A powerful 50-pound pull magnet

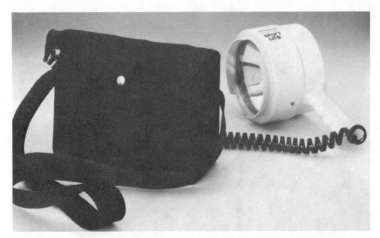

Fig. 5-4. Q-Beam Yellow Sportlite and Q-Beam Power Pack (courtesy Brinkmann Corporation).

allows the light to be positioned for hands-free use. Adjustable positioning allows 180-degree vertical swivel of the light which can also be removed for hand-held use. A two-position switch allows a spot light beam of 200,000 candlepower or a flood light of 100,000

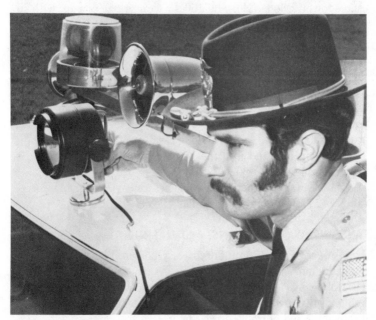

Fig. 5-5. The Q-Beam Magnetic Base Spot/Flood light mounted on top of vehicle (courtesy Brinkmann Corporation).

Fig. 5-6. Q-Beam battery pack with shoulder strap and recharger.

candlepower. Model No. 800-1501-0 weighs 3¾ pounds with its magnetic base and 15 feet of straight power cord and lighter plug.

Q-Beam Rechargeable Battery

The portable rechargeable power pack provides 12 Vdc for use with the various Q-Beam Lights. You can use it with your Q-Beam when on foot or for on-site inspections away from your vehicle. The power pack shown in Fig. 5-6 comes complete with ac recharger adapter and shoulder strap carrying case. A quartz halogen light is shown being inserted into the battery pack in Fig. 5-7. The 9-ampere model is No. 802-1722-1.

The Q-Beam products are manufactured by Brinkmann Corporation, 4215 McEwen Road, Dallas, Texas 75234.

LANTERNS, WARNING LIGHTS, AND FLASHLIGHTS

An estimated 170 million battery-powered lights are now in use, with more than 30 million of them being replaced every year. The following sections will familiarize you with some of these lights, from the world's smallest rechargeable flashlight to one-year disposables and portable strobe lights.

Electronic Flare

The Electronic Flare is a dependable emergency signal which flashes a bright red beam for over 20 hours at the rate of one flash per second. A red lens provides a brilliant red beam from a disposable three-cell, 4½-volt battery. The unit is waterproof and shock resistant. Model No. 273 is shown on the right in Fig. 5-8. It measures 3 × 1½ × 4½ inches.

Life Lite

This is one of the brightest, most dependable disposable flashlights available. Shown on the left in Fig. 5-8 with a white light, it will shine a 300-foot power beam. The unit is powered by a three-cell 4½-volt battery and is available in colors of red, white, blue, yellow, and green. It measures the same as the Electronic Flare and is shown in use in Fig. 5-9 by a fireman where the light is inserted in the strap on his hat and will shine wherever he turns his head.

The Electronic Flare and Life Lite are manufactured by Garrity Industries, 14 New Road, Madison, Connecticut 06443.

Brinkmann 6-in-1 Troubleshooter

There are a number of features combined in the 6-in-1 Troubleshooter. Fig. 5-10 shows the front of the unit with the amber

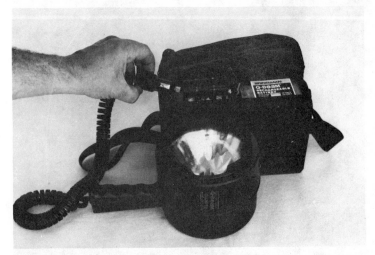

Fig. 5-7. Lamp being inserted into Q-Beam battery pack.

Fig. 5-8. The Electronic Flare and Life Lite by Garrity Industries. The electronic flare on the right flashes a red light.

light on. The dark light at the bottom is the red flasher light that alternates with the amber light. The 6-in-1 has the following features.

- Simultaneous operation of the amber and red warning

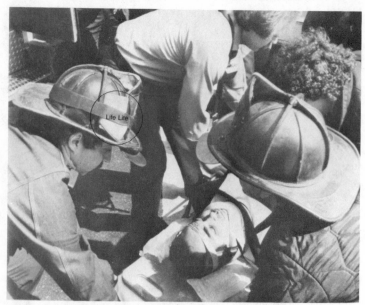

Fig. 5-9. Life Lite by Garrity Industries inserted in fireman's hat.

Fig. 5-10. Brinkmann 6-in-1 Troubleshooter light (courtesy Brinkmann Corporation).

flasher and the fluorescent lantern which comes with the 6-in-1 warns on-coming traffic and provides bright illumination of the working area.

● Powerful spotlight/searchlight penetrates darkness to illuminate signs, road hazards, shoreline, with 9 volts of power.

- Brilliant fluorescent lantern lights up camping, fishing, working area with 6-watt, 6000-hour rated tube. Shatterproof lens guard.
- Blinking amber/red light flashes warning when caution is needed, day or night. Diamond-etched shatterproof lens.
- Steady red emergency light warns of danger, attracts attention in any emergency situation.
- Six-position rotary switch for instant choice of light setting.
- Operates on six D-cell batteries.
- Idea lightweight all-purpose light for automotive, boating, fishing, hunting, utility emergency, and camping use.

In Fig. 5-11, we see the fluorescent light in operation with the red and amber flashing lights. The 6-in-1 Trouble Shooter is manufactured by Brinkmann Corporation, 4215 McEwen Road, Dallas, Texas, 75234.

Rechargeable Lantern

This rechargeable lantern, Model 150 by Nicholl Brothers, can

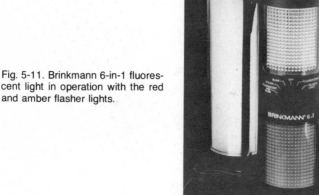

Fig. 5-11. Brinkmann 6-in-1 fluorescent light in operation with the red and amber flasher lights.

Fig. 5-12. Nicholl Brothers Challenger Lantern Model 150.

be recharged up to 200 times using internal nickel cadmium batteries. Shown in Fig. 5-12, this lantern will provide up to two hours of light on a single charge which would cost about a penny. The 120-Vac charging system is self-contained and comes complete with 6-foot cord and ac plug. The light uses an adjustable focus and has a 120-degree swivel stand. Table 5-1 shows a use/charging table for the Model 150. For maximum battery life the lantern should be left plugged in on charge whenever it is not in use as it cannot be overcharged.

Specifications on the Nicholl Model 150 Challenger Lantern are shown in Table 5-2. This unit is manufactured by Nicholl Brothers, Incorporated, 1204 West 27th Street, Kansas City, Missouri 64108.

Table 5-1. Use Versus Charging Time for Model 150.

Time in Use	Hours of Charging to Fully Charge
One-half hour	8 hours
One hour	16 hours
Over one hour	24 hours

Table 5-2. Model 150 Challenger Specifications.

Item	Description
Lens/reflector	Sealed, focusing, dual reflector
Battery system	120 Vac rechargeable NiCads
Bulb	Precision focused PR2
Switch	Sealed rotary w/plated contacts
Case	Rain tight, high impact, burgundy color
Stand	120 degree swivel, chrome-plated steel
Handle	Full-sized, with glove clearance
Size	4 15/16 × 5½ × 8 7/16 inches
Weight	1 pound 8 ounces

Floating Lantern Model 511

The Dark Blazer All-weather Floating Lantern, Model 511, features a Super II reflector system that produces 77 percent more candlepower for a much brighter beam. The Model 511 is shown in Fig. 5-13 sitting on a 120-degree swivel stand which locks up and

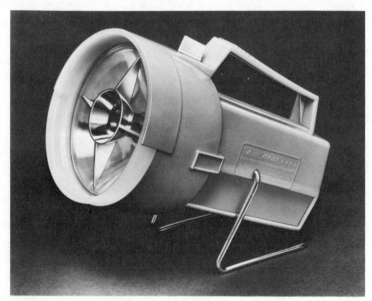

Fig. 5-13. Nicholl Floating Lantern Model 511.

out of the way when not in use. The stand is also removable. The 511 is made of high-impact plastic and is colored bright red. The lantern has a 4½-inch reflector with adjustable focus and a place for spare bulbs.

The Model 511 uses a standard 6-volt spring-terminal battery and has a four-way switch which gives the user a choice of Off, Beam Only, Flasher Only, and Beam and Flasher. An interior warning flasher illuminates the front part of the red case for emergencies.

The Floating Lantern Model 511 is manufactured by Nicholl Brothers, Incorporated, 1204 West 27th Street, Kansas City, Missouri, 64108.

Headlights by Brighteyes

When we think of headlights, we immediately think of an automobile. But these lights by Brighteyes are worn on the head—hence, Headlights! This uniquely designed flashlight fits like a pair of glasses, and yet can be worn with or without your regular glasses. There are two powerful lights on each side which shine a bright beam of light wherever you look. When doing hard-to-get-at work, Headlights will light the way as shown in Fig. 5-14. Headlights are a marvelous aid for the handyman, as they give him complete freedom to use his hands for tools and equipment while the light illuminates the work area.

A switch is built into the comfortable polypropylene frame so

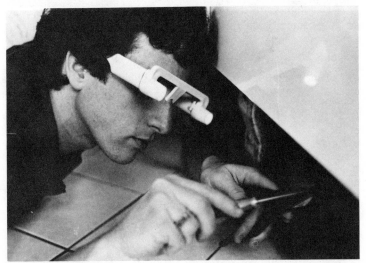

Fig. 5-14. Headlights are worn just like glasses, leaving hands free for work.

Fig. 5-15. Headlights in use at a camping site (courtesy Brighteyes).

that the device shuts off automatically when you fold the collapsible earpieces. The light is powered by four AAA alkaline batteries which are inserted into the plastic frames, two batteries on each side.

In Fig. 5-15, we see the Headlights in use at night on a camping trip. But they can also be used at home, on the farm, or while bicycling, flying, boating, fishing, hunting, backpacking, mountain climbing, and making electronic repairs.

Headlights are manufactured by Brighteyes, 1815 East Carnegie Avenue, Santa Ana, California, 92705.

Handi-Brite

This is another hands-free type flashlight which can be worn on the head or about the body using appropriate straps. The Handi-Brite is shown in Fig. 5-16 where a headband keeps the flashlight in place. There are hundreds of uses for the light which uses two C cells and a broad-beam reflector for bright, wide light. The Handi-Brite has a swivel front so you can tilt the reflector up or down 54 degrees to aim the beam exactly where it is needed.

In addition to wearing the Handi-Brite on your head by means of a comfortable headband, you can also wear it on your belt with a wide secure clip which fastens to belts or wastebands. The Handi-Brite will clip to your car visor for reading maps or during any auto emergency. You can also hang the light around your neck like a pendant on a supplied cord and swivel the light to any desired position. You can rest it on your chest while in bed so you can read without disturbing others.

A bracket is supplied with the Handi-Brite so you can fasten it to any wall. Rest it on any surface to shine on work areas under sinks, dashboards, and the like. You can also use the light at night for jogging, bicycling, camping, backpacking, hunting, and boating. The light is handy for emergencies such as power failures, auto breakdowns, and fire evacuation. Handi-Brite is ideal for patient or baby care where you can administer medicine or change diapers without harsh room lights. Because of its rugged construction of

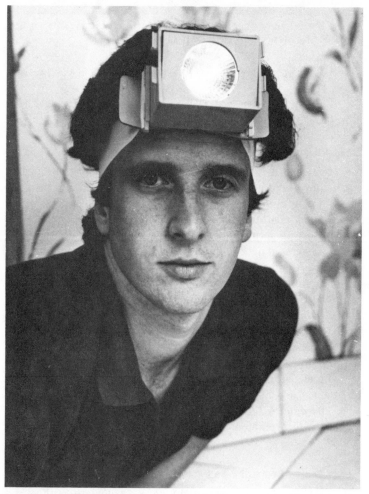

Fig. 5-16. The Handi-Brite light is worn on the head for hands-free work.

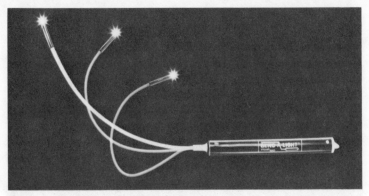

Fig. 5-17. The Bend-A-Light can be shaped into many positions for small-hole inspection (courtesy L&W Enterprises).

ABS plastic, copper wire, and brass connections, it can be used underwater.

This unusual light is available from Handi-Brite Incorporated, Box 1735, Hains City, Florida 33844.

Bend-A-Light

The Bend-A-Light is a revolutionary flexible light tool which is a must for every home, office, hobby, shop, and auto. The light, shown in Fig. 5-17, can be bent and rebent in any desired shape to get into any quarter-inch hole to provide illumination impossible with ordinary lighting. The Bend-A-Light is a high intensity, pin-point, prefocused light on a 10-inch, flexible shaft one-eighth inch in diameter. A tiny optical lens produces a brilliant beam of light that does not reflect into the eyes.

The Bend-A-Light unit is made of nonconducting material so shock hazards are eliminated. An attachable mirror, magnetic pick-up, and clip-on magnetic holder for hands-free operation are also available. The light is powered by two AA batteries. Figure 5-18 shows No. B030 Bend-A-Light Versatool which is complete with extension/cover, illuminating pick-up magnet, heavy duty clip-on magnet, inspection light, and extension mirror attachment.

The Bend-A-Light Visual Tool is manufactured by L&W Enterprises, Incorporated, 200 S. Washington, Royal Oak, Missouri, 48067.

Sears Sealed Beam and Red Flashing Light

This sealed beam flashlight comes complete with a red

emergency flashing light. The whole unit mounts on top of a standard 6-volt screw terminal battery, as shown in Fig. 5-19. The flashing action of the red blinker light is achieved by a self-flashing (bi-metal) bulb. Two separate switches are used to control the

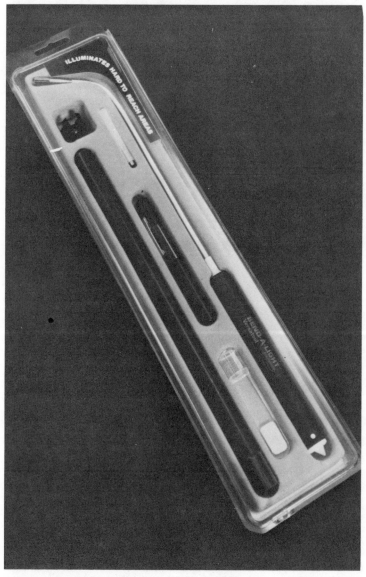

Fig. 5-18. Bend-A-Light Model No. B030 packaged with accessories (courtesy L&W Enterprises).

Fig. 5-19. Sears sealed beam flashlight and red blinker.

action of the sealed beam flashlight and the red flasher. Means of attaching the sealed beam and flasher unit to the screw terminal battery is shown in Fig. 5-20. The position of the sealed beam light is adjustable so the light can be set on a surface and the light positioned to any desired angle.

The Sears Sealed Beam and Red Flasher Light is available from most Sears stores.

Sears Rechargeable Flashlight

This rechargeable flashlight will give about 25 percent more light than a standard two-cell flashlight equipped with D cells because of the special bulb used. It will operate for approximately two hours before recharging is required. Figure 5-21 shows the Sears rechargeable flashlight plugged into a 120-Vac wall outlet for recharging by using the built-in charger. Because it uses NiCad batteries, it can be recharged approximately 200 times. Table 5-3 shows charging time required after light usage.

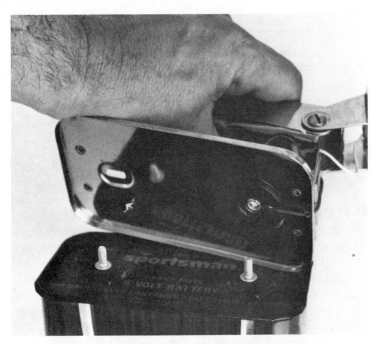

Fig. 5-20. Sears Sealed Beam unit being attached to a 6-volt screw terminal battery.

Fig. 5-21. Sears Rechargeable Flashlight being recharged from 120-Vac wall outlet.

93

Table 5-3. Charging Time for Sears Flashlight.

Time in Use	Hours of Charge for Full Charge
30 minutes	6-8 hours
1 hour	12-15 hours
1½ hours	18-20 hours
2 hours	24-30 hours

This rechargeable flashlight is available from your local Sears stores.

Black and Decker Spotliter

The Spotliter is a rechargeable light that stores in any room in the house in its own storage unit so that it is fully charged when you need it. Figure 5-22 shows the Spotliter, the recharging unit which plugs into 120 V ac, and the storage unit which can be mounted on a wall with the supplied screws. The three NiCad rechargeable cells eliminate corrosion problems of ordinary batteries and supply enough power to provide light for over one-and-a-half hours. The Spotliter is made of rugged ABS plastic and has a 6-foot cord and a hidden cord wrap. The flashlight is almond colored and has a two-

Fig. 5-22. Black and Decker Spotliter with storage base and charger (courtesy Black and Decker).

Fig. 5-23. The Endura 365 One year Flashlight by Wonder Corporation (courtesy Wonder Corporation).

position switch for regular or a burst of power. The charger storage base is brown colored. A red LED on top of the unit glows to indicate that the unit is plugged in and is being charged. The whole unit weighs 1.8 pounds.

The Spotliter is Model 9360 and is manufactured by Black and Decker, Incorporated, Towson, Maryland 21204.

Endura 365 The One Year Light

This dependable, disposable flashlight is made to produce 4½ volts of power for the light. The unit has a luminous switch for locating in the dark and is shown in Fig. 5-23. It is made in colors of green, red, blue, yellow, and ivory. The 4½ volt power source is equivalent to three C cells.

The Endura 365 is manufactured by Wonder Corporation of America, Norwalk, Connecticut, 06856, and is available in many stores.

Key-Lite

The Key-Lite is a handy key-chain disposable flashlight and is compact for purse or pocket. The light is shown in Fig. 5-24 and is equipped with a chain which can be used to keep keys in order. The light has a prefocused bulb and a switch on the end of the key lead is used to turn the light on and off. This type of key light is convenient for auto, home, and office use.

The Key-Lite is manufactured by Wonder Corporation of America, Norwalk, Connecticut 06856.

Deluxe Penlite

This is a high intensity light that is sealed for longer life. As shown in Fig. 5-25, the pocket clip is pressed to turn on the light. The unit is made of a brushed metal case and operates off two AAA

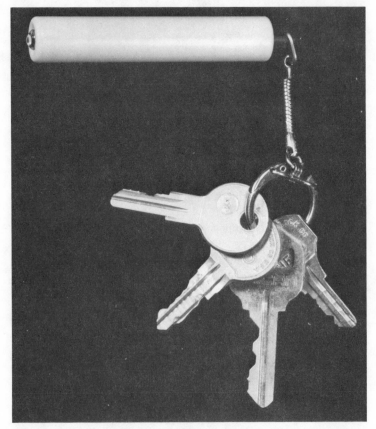

Fig. 5-24. Key-Lite by Wonder is a high intensity flashlight with keyring.

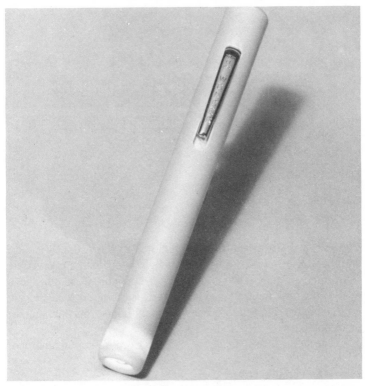

Fig. 5-25. The Deluxe Penlite is made by Wonder and can be carried in the pocket (courtesy Wonder Corporation of America).

cells which are replaceable. The Penlite is a high intensity disposable light also made by Wonder.

The Deluxe Penlite is manufactured by Wonder Corporation of America, Norwalk, Connecticut 06856.

High-Intensity Disposable Flashlight

The Dalon is a high-intensity focused beam flashlight which is disposable and sealed for longer life. Figure 5-26 shows the Dalon with a side-mounted switch. The unit is convenient for pocket, purse, glove compartment, or use anywhere. It comes in colors of blue, red, yellow, and green.

The Dalon high intensity disposable flashlight is made by Wonder Corporation of America.

Ultra-Lite

The Ultra-Lite has the light output of a three-cell flashlight

Fig. 5-26. The Dalon Disposable flashlight by Wonder (courtesy Wonder Corporation of America).

with a 4½-volt disposable battery. Figure 5-27 shows the Ultra-Lite with oversized reflector for brighter light which puts out a visible beam up to 500 feet. The unit is water and moisture resistant and is equipped with a multiple-use stand and hang bracket which easily attaches to belt or pack. It is convenient for home, car, boat, purse, or camping.

The Ultra-Lite is made by Wonder Corporation of America.

Twist-Top Portable Light

This is a portable light with a twist top switch so that you can stand it, hang it, or carry it. Figure 5-28 shows the wide area lighting top of the Twist-Top light which operates off four D cells.

Fig. 5-27. Ultra-Lite flashlight by Wonder has oversized reflector.

Fig. 5-28. The Twist-Top light is made by Wonder and operates off four D cells (courtesy Wonder Corporation of America).

The unit is great for camping, boating, and auto, or home lighting during power outages.

The Twist-Top Light is made by Wonder Corporation of America.

Swivel Lite

This high intensity light features a suction cup stand for use at home, auto, or office. Fig. 5-29 shows the suction cup and light which operates off two AA cells. The suction cup can also be removed from the light so it can be used as an ordinary high

Fig. 5-29. The Swivel Lite has a suction cup mount and swivel for easy movement (courtesy Wonder Corporation of America).

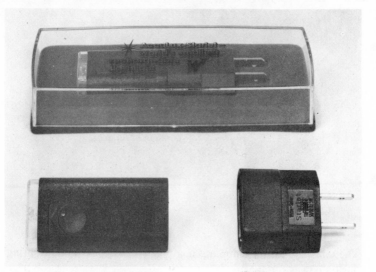

Fig. 5-30. The Starlet is described as the world's smallest rechargeable flashlight (courtesy Wonder Corporation of America).

intensity light. The Swivel Lite is made by Wonder Corporation of America.

Starlet

The Starlet is the world's smallest rechargeable flashlight and is shown in Fig. 5-30. A prefocused bulb with lens will provide 40 minutes of light on one charge and it costs just pennies to recharge. As shown, an adapter is used to connect the rechargeable batteries to 120 Vac so that it can be plugged into an ac outlet.

The Starlet rechargeable flashlight is made by Wonder Corporation of America.

Astrolite Five-Function Compact Lantern

This combination flashlight and fluorescent lantern performs five functions in all. It is a perfect mini-lantern for personal use in auto, home, office, emergency, and travel. Pictured in Fig. 5-31, the Astrolite operates on four C cells and provides a wide area fluorescent light as well as a powerful incandescent spotlight. There is also an emergency steady amber warning light and a warning amber transistorized blinker light. Finally, there is a red passive reflector which is illuminated by on-coming auto headlights. A handy strap is provided for easy carrying or hanging on a nearby support at camp.

Fig. 5-31. The five-function compact Astrolite lantern (courtesy Wonder Corporation of America).

The Astrolite is made by Wonder Corporation of America.

Road Runner Safety-Lite

The Road Runner light is especially useful when jogging, walking, and cycling at night. The flashlight has an all-weather adjustable armstrap as shown in Fig. 5-32. The light provides 180

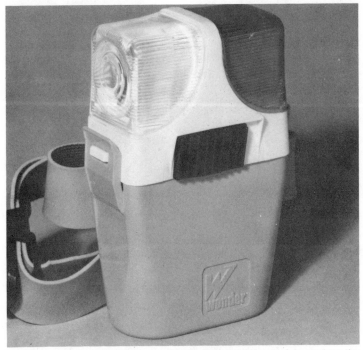

Fig. 5-32. Joggers will like the Road Runner Safety-Lite which is worn on the upper arm (courtesy Wonder Corporation of America).

degrees of visibility and has a white lens that faces forward of the runner and a red lens that faces to the rear of the runner. The light is compact, lightweight, and comfortable. The light operates off two C cells and has a positive easy-to-reach switch.

The Road Runner Safety-Lite is made by Wonder Corporation of America.

Bike Safety-Lite

This light provides three-way visibility for night-time safety for bikers. Figure 5-33 shows the Bike Safety-Lite which has a reflective and translucent red lens facing the rear and a white lens facing forward for maximum visibility. An amber dome light provides side visibility for the biker who can wear the light on his upper

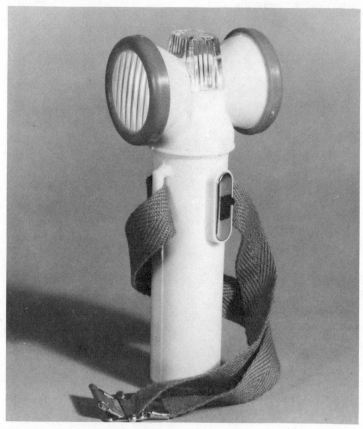

Fig. 5-33. Bikers will like this three-way Bike Safety-Lite (courtesy Wonder Corporation of America).

arm or leg as the light is provided with an adjustable strap. The unit is lightweight and is provided with a positive on/off switch. The Bike Safety-Lite is powered by two C cells and will provide hours of light safety for nighttime use.

The Bike Safety-Lite is made by Wonder Corporation of America.

Durabeam Flashlights and Lanterns

Duracell USA, manufacturers of the copper top battery, the Duracell®, has introduced a family of five Durabeam® flashlights as shown in Fig. 5-34. In the lower left of the figure we see the Durabeam flashlight powered by two alkaline D-cells. The Durabeam flashlight is constructed of the same durable plastic used in the helmets of pro football players and it has a switching system—tested for 50,000 on-off cycles—that is guaranteed for the life of the flashlight. The unit is weatherproof, will not corrode or rust, and will still work after being dropped on a concrete floor at 0° F.

In the lower right of Fig. 5-34, we see the Compact light powered by two AA alkaline cells, which is also made of durable plastic and will continue to work after being dropped on concrete. In the center of Fig. 5-34, the Durabeam lantern is pictured which is powered by one 6-volt alkaline lantern battery. The lantern will float and has a switch also tested for 50,000 on-off cycles. It will also not rust or corrode and will work after being dropped on concrete at 0° F.

Pictured in the upper left of Fig. 5-34 is the Durabeam emergency light which is powered by one 6-volt alkaline battery. A feature of this light is a unique self-adjusting spot beam head that keeps its aimed position without exposing wiring and circuitry to damage. It also has a flasher beam in the handle.

In the upper right of Fig. 5-34 we see the Durabeam area light powered by one 6-volt Duracell alkaline battery. This area light also has a two-year warranty and delivers twice as much light as other generally available area lights. The light also has a pop-out hanger ring and is easy to pack for storage and transportation.

Strobolite

The Strobolite is an electronic flashing signal light made by Honeywell and is handy for any emergency situation that might arise. The flashing light is clearly visible from many miles away in any direction. The unit is pocket-size, waterproof, and weighs just

Fig. 5-34. Durabeam flashlights and lanterns by Duracell (courtesy Duracell USA)

11 ounces with batteries. Figure 5-35 shows the Strobolite with its convenient wrist lanyard which allows it to be hung from tree branches or a nail in the garage.

The Strobolite can be used for cross-country skiing, for hiking and camping, for boating and scuba diving, or hunting and fishing trips, when motorcycling, or when snowmobiling. The unit is compact, measuring just 6 × 2½ × 1⅝ inches. It will flash for up to seven hours when equipped with a fresh set of batteries. The Strobolite is designed to operate between 0 and 122° F. It operates about seven hours at 60-70° F, one-and-a-half hours at 0° F, and four to five hours at 122° F.

The Strobolite provides a brilliant omnidirectional beam, visible for many miles, and is 300 times brighter than an ordinary flashlight. The unit operates off ordinary alkaline C cells and there is no bulb or filament to burn out. Strobolites of this type are ordinarily good for about a million flashes. The housing is waterproof and noncorrosive. The light will even float while flashing so it is useful while boating or fishing.

The nominal flashing rate is between 70 and 30 flashes per minute, depending on operating temperature and battery condition. Because it is a pulsed strobe light, it provides cool electronic light and can be hand-held anytime. It can be used for communication, location, and emergency use. A Strobolens is available for use on the Strobolite which will produce emergency red, amber, or blue high-visibility colors.

Fig. 5-35. The Honeywell Strobolite flashes a brilliant light visible for miles. The electronic flash is 300 times brighter than a flashlight (courtesy ACR Electronics, Incorporated).

The Strobolite is Honeywell Catalog No. 2700 and is available from ACR Electronics, Incorporated, P.O. Box 2148, Hollywood, Florida 33022.

INNOVATIVE PROJECTS USING A FLASHER LED

For the do-it-yourselfers, the circuits that follow will give the experimenter an opportunity to try various circuits which provide for a flashing light emitting diode (LED). The flasher LED is available from many sources and is carried by Radio Shack as Catalog No. 276-036. The unit is inexpensive and offers some very interesting possibilities for circuit innovation.

The flasher LED has its own built-in IC switch which flashes it at a rate of about three flashes per second when operated off 5 Vdc. We will describe many novel configurations which use a 9-volt battery, a photocell, and resistor. An LED flasher can be used for a night light, a basic flasher for TTL (transistor-transistor-logic) and CMOS (complementary metal oxide semiconductor) circuit applications, attention-getter applications, a trouble-shooting aid, security warning light, and ambient light or dark detector.

Basic Circuit Operation

The basic LED is made to flash at a three-times-per-second (PPS) rate by a small integrated circuit that operates off 5-Vdc power. The flasher LED is unique when you consider that the whole LED is the same size as a regular LED but contains electronic circuitry that includes the following electrical components: the LED, the IC chip that establishes the flash rate (in effect containing an RC time constant circuit to switch the current to the LED), and an effective resistor that drops the supply voltage from 5 Vdc to a nominal 1.6 volts for application to the red LED.

The flasher draws 20 mA from a 5-volt source, so for a red LED which draws 20 mA, a series-dropping (current limiting) resistor of 170 ohms would be required. Therefore, a lot of action is accomplished by the small IC chip that can be seen as a small black speck inside the LED epoxy case. At the present time, the flasher is available only in red, but other colors—such as green and yellow—probably will be forthcoming from the various LED manufacturers.

Circuit Applications

Some circuit applications that make use of the flasher LED are described in the sections that follow. These are all fairly simple circuits and can easily be completed in a few minutes using alligator

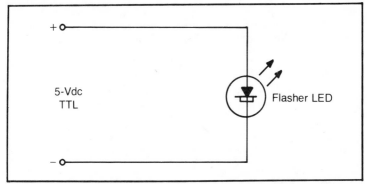

Fig. 5-36. LED flasher is powered from 5-volt supply and flashes at a 3 PPS rate.

clips for a quick hookup to see if it is the circuit you want.

Normal Flash Rate. In Fig. 5-36 we see the basic circuit hookup for the LED flasher operating off 5 Vdc. This is a standard circuit configuration that would be used to drive the flasher directly off the 5 Vdc used in TTL and CMOS circuits. It will flash at a nominal 3 PPS. In Fig. 5-37 we see an arrangement for flashing the LED from a 9-volt transistor battery at a 3 PPS rate. A 9-volt battery will provide sufficient power to flash the LED for about a week or so, and so for continuous operation, you can use a 9-volt battery eliminator or charger available from local radio supply stores for under $5.

An alternating current power supply of a nominal 6.3 to 9 Vac can be used to power the flasher by adding a diode to the circuit to protect the LED IC chip during negative voltage swings of the ac voltage. This circuit arrangement is shown in Fig. 5-38. The 6.3 to 9

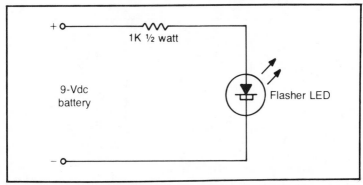

Fig. 5-37. Power for flasher LED can be obtained from a 9-volt source if a dropping resistor is used.

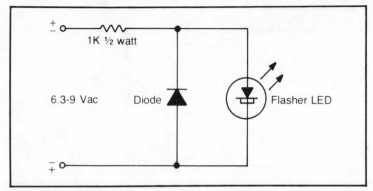

Fig. 5-38. Ac power can be used if a diode and dropping resistor are included in circuit.

Vac power supply can be obtained by using a 115-volt to 6.3-volt filament transformer or an ac pocket calculator charger which usually has a nominal 8-Vac output.

Fast Flash Rate. We can increase the flash rate by adding a large capacitor across the series-dropping resistor as shown in Fig. 5-39. The flash rate is increased to a nominal 10 PPS by the RC circuit introduced in series with the IC chip. The capacitor can be any value from 500 to 3000 microfarad at a nominal 10 to 35 Vdc working voltage. Experiment with the value of R and C until you reach the flash rate you want.

If the flash rate is increased to slightly above 10 to 12 PPS, the LED will appear to be on continuously as the eye cannot perceive faster flash rates. To observe the LED flashing (if your circuit leads are long enough), wave the LED back and forth slowly and you will see it make on-off streaks as it moves.

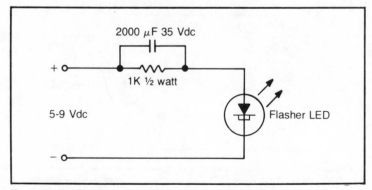

Fig. 5-39. Flash rate can be varied with a large capacitor RC circuit.

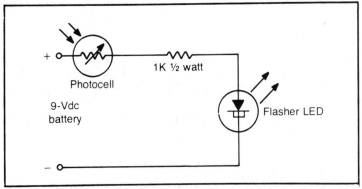

Fig. 5-40. A series photocell permits LED to flash only in bright ambient light conditions.

Ambient Light Detector. When we put a photocell in series with the flasher LED as shown in Fig. 5-40, it will flash only in the presence of light. Photocells are available from any radio supply store and have a nominal resistance of 1 to 10 megohms in darkness and their resistance drops rapidly to a nominal 100 to 1000 ohms in bright light. In darkness, the circuit will draw virtually no standby power as the total resistance in the circuit is over 1 megohm. Considering the IC chip as a short at this time, the circuit will draw only 9 μA from the battery, virtually that of its shelf life. You can use this circuit to tell you when it gets dark outside (if you are in a windowless room) or if you *really* want to see if the refrigerator light goes out when the door is closed!

For light levels between light and dark, where the applied voltage to the IC chip will vary from 0 to 5 volts, we will find the flasher LED doing some strange things such as flashing faster, slower, staying on or off, and varying in brilliance.

When we place the photocell across the flasher LED as shown in Fig. 5-41, we now find that the LED will not flash in bright light (the low resistance of the photo cell shorts out the IC chip) but when the photocell is in darkness, the LED will flash. In darkness, the photocell resistance rises to about 10 megohms; this appears as an open circuit to the IC chip. Five volts appear across the chip and the LED begins to flash. This circuit will draw power from the battery in the standby (light present) condition and nominal power when flashing, so you might want to use a 9-volt battery eliminator for long-term operation. This circuit is handy for a flashing night light or in areas where you might need to know that a certain light is still on and operational.

109

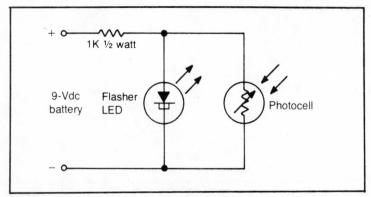

Fig. 5-41. A parallel photocell permits the LED to flash only in dark ambient light conditions.

Alternate Flashing Red and Green LEDS. The flasher LED can be used in a circuit arrangement as shown in Fig. 5-42 to alternately flash a second LED. The two LEDs can be spaced several inches to several feet apart to attract your eye back and forth to each LED as it flashes. Alternately flashing red and green LEDs are particularly interesting as they are eye-catching and can serve as baby-sitters, novelties, or attention getters. The circuit of Fig. 5-42 will operate from a nominal 3 to 6 Vdc, the flash rate increasing as the voltage is decreased. At 6 Vdc, such as you get from a Type F

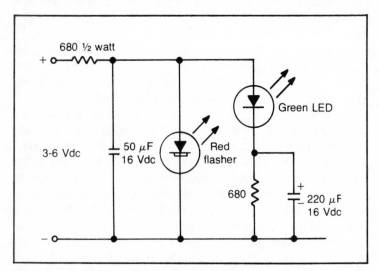

Fig. 5-42. Red and green LEDs flash alternately at a rate determined by the applied voltage.

lantern battery, the flash rate is the nominal 3 PPS. If the circuit voltage is increased past 6 volts, up to 7 or 8 volts, the LEDs will stop flashing and remain on continuously. That condition should be avoided, for long-term use might damage the IC chip in the flasher LED.

As the voltage is reduced to about 3 volts, the flash rate increases to about 10 PPS and the LEDs are not as bright as at 6 volts. The LEDs will flash faster and faster as the voltage is reduced below 3 volts until they appear to be on continuously, though they are dim at this fast flash rate.

As you experiment with the flasher LED, you will find it a very interesting electronic component. You may observe that its flash rate changes, depending on the amount of ambient light striking the IC chip inside its epoxy case. Depending on the manufacturer of the LED, the flash rate will be a nominal 3 PPS in bright bench light or sunlight. But as you darken the room, the flash rate will decrease slightly, depending on the circuit you are using at the time. Do your own experimenting with this unique LED until the manufacturers correct for some of its interesting characteristics. They might add a zener voltage regulator to keep the flash-rate constant with applied voltage and then hide the IC chip in a light-tight case.

LED FLASHER IC CHIP

The LM3909 Flasher/Oscillator IC chip is ideal for designing small, battery-powered flashers that will last for many months or years. The LM3909 will generate enough voltage to flash an LED working with battery voltages as low as 1.1 volts. As a matter of fact, you can use your old batteries you are about to discard and power this flasher unit for a number of months. In such low-duty cycle applications, batteries will appear to last indefinitely.

Small-Count IC Chip

The LM3909 LED flasher is probably the smallest component-count IC chip designed. With just a single capacitor added, the IC will flash an LED when the chip is powered by a 1½-volt battery. Fig. 5-43 shows the internal circuit diagram for the LM3909 and the external wiring required for the flasher. The IC chip and components can be mounted on a single IC board with leads extended for soldering and breadboarding. Such an IC board is available from Radio Shack as No. 276-024 which will mount one 6 to 16-pin IC chip. This board can be cut in half to provide two boards of 8 pins each since the LM3909 is an 8-pin DIP package.

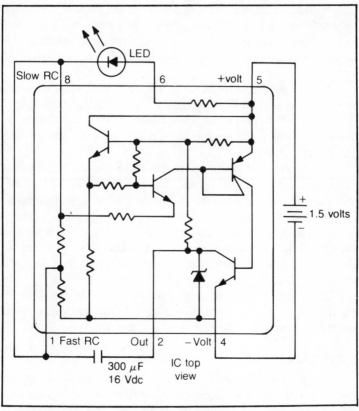

Fig. 5-43. Timing and voltage boosting functions are provided by a single capacitor in the LM3909 LED flasher circuit design.

Originally designed as an LED flasher, the LM3909 also provides an ideal trigger for silicon-controlled rectifiers (SCR) and triacs. It will also flash small incandescent lamps. Its ability to operate with only 1½-volt battery supplies gives the LM3909 several rather unique characteristics. For one, no known connection can cause immediate destruction, meaning you can experiment and not damage the chip, the second, its internal feedback loop insures self-starting of the properly loaded oscillator circuits.

Uses for the Flasher

The LM3909 flasher is handy for flashing a low-level light source to be used in a child's room for frame of reference in case of a power outage during the night. Placed in a window, it will cause an intruder to wonder if the house is wired for security with a blinking, timed sensor installed in each window. You can also place one on the dashboard of your auto so that it appears your auto is wired for theft protection and the system is "armed." You can also consider making novel jewelry so that a brooch flashes a red or green LED.

Circuit Description

Referring to Fig. 5-43, minimum power dissipation is achieved in two ways. Operationally, the LED draws current approximately 1 percent of the time. During the remaining time, all transistors except Q4 are Off and the drain through the 20 k resistor in Q4's emitter is only 50 μA. The charge path for the 300 microfarad capacitor is through the two 400 ohm resistors connected to pin 5 and the 3 k resistor connected to pin 4.

Transistor Q1 and Q3 remain Off until the capacitor has charged to about 1 volt. The junction drop of Q4, its base-emitter voltage divider, and the junction drop of Q1 determines this voltage level. When the voltage at pin 1 is a volt or more negative than the level at pin 5, Q1 begins to conduct, turning on Q2 and Q3, supplying a high current pulse to the LED.

The Q2-Q3 transistor combination has a current gain of between 200 and 1000. Q3 can handle over 100 mA and rapidly pulls pin 2 close to the supply voltage common (pin 4). The capacitor at pin 1 goes rapidly below supply common. The voltage available for the LED is thus higher than the battery voltage. The 12-ohm resistor between pins 5 and 6 limits the LED current to a safe value.

The LM3909 is indeed an interesting IC chip flasher and will provide many years of operation, especially if a larger capacity cell is provided. The LM3909 will operate off 1½ to 6 Vdc, though its main claim to fame is low voltage, low power operation.

NEON BLINKER

The neon blinker is designed to operate off 6 Vdc and uses a transistor to switch current through a transformer to boost the voltage to a level sufficient to flash a neon bulb. The circuit shown in Fig. 5-44 also uses a minimum number of parts to form an oscillator circuit. The circuit boosts the 6 volts to 67 Vdc, the level required to flash a neon bulb.

The heart of the circuit is the unusual output transformer which boosts the voltage. The rate of flashing the neon bulb is largely determined by the setting of R1, the 2 k potentiometer. You will note that there is no on-off power switch in the circuit. Because the circuit draws so little power to flash the neon bulb, there is no need

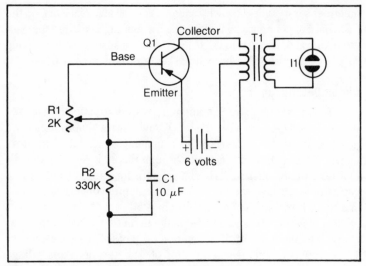

Fig. 5-44. Schematic diagram for neon light blinker.

for a switch! It will flash almost forever on four size D flashlight cells.

Table 5-4 shows the parts required for the neon light blinker. These parts are available at your local electronics parts stores.

DUAL-PURPOSE SAFETY FLASHER

The circuit for the dual-purpose safety flasher shown in Fig. 5-45 can be used anywhere a hazard or potential hazard exists. With the switch set on Auto, oncoming headlights will illuminate the photocell and trigger the circuit so that the No. 257 lamp starts flashing. When placed in the Continuous mode, the light will flash all the time, independently of any oncoming traffic. The switch should

Table 5-4. Parts List for Neon Light Blinker.

Item	Description
B1	6-volt battery (4 D cells)
C1	10 μF, 10-volt electrolytic capacitor
I1	NE-2H neon bulb
Q1	PNP transistor PN 2484
R1	2K potentiometer
R2	330K resistor 10 percent ½ watt
T1	Universal output transformer
Misc.	Wire, battery holder, solder

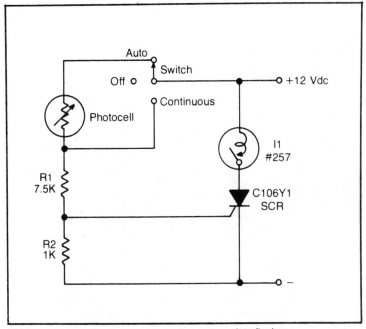

Fig. 5-45. Schematic diagram for dual-purpose safety flasher.

be set to Auto in a disabled car parked on a lonely road to conserve battery power. On a heavily travelled road or highway where traffic is flowing at high speed, it is safest to put the switch in the continuous position to provide long-distance warning to on-coming traffic. You can also use the flasher wherever you want to warn motorists and pedestrians of open-pit excavation by operating the unit in the Continuous mode.

The parts list is shown in Table 5-5 and most parts should be available at any local electronics supply store.

Table 5-5. Parts List for Safety Flasher.

Item	Description
I1	Lamp #257
PC	Photocell, cadmium sulfide (CdS)
R1	Resistor, 7.5K, 10 percent, ½ watt
R2	Resistor, 1K, 10 percent, ½ watt
SCR	Silicon-controlled rectifier, C106Y1
SW	Switch, DPDT, center off
Misc.	Wire, solder

6 | Home and Camp Power Generation

Storms, accidents, or failure of power-supplying equipment at times can cause power outages. An outage that lasts for a long period can lead to serious problems such as frozen water pipes, spoiled food, household or camping accidents, and the like. However, with standby electric power service, you can keep your essential electric equipment operating during power outages. This "insurance" can help you avoid inconvenience and possible financial loss. To provide standby service, you need an alternator (generator) to produce 60 Hz (cycles per second) of alternating electric current at 120 or 240 volts, an engine to run the alternator, and a transfer switch to safely control the electric current.

In this chapter we will discuss items to help you be more knowledgeable in the selection of emergency power generating equipment. We will also cover some commercial power devices which are available. In this way, you can get an idea of what you might need to satisfy your requirement for emergency or camping power generation.

BUYING AN AC POWER GENERATOR

A portable ac generator is a positive answer to the need for powering electrical tools and devices during a power outage or where a power line is not available. It can supply 120-volt, 60-Hz ac power to operate standard electrical items in the home, backyard, or on the "back 40." That is why it has been such a popular item for powering not only a host of shop and garden tools, but also TV sets, radios, steam irons, hair dryers, mixers, toasters, blenders, vacuum cleaners, refrigerators, lighting, and many more of the appliances that enhance our enjoyment of modern living.

A portable generator breaks the umbilical attachment of these devices to the home power outlet: you *can* take 'em with you, because you have a way to feed them the electrical power they need! But there is another fact of life to consider when you are looking at portable ac generators: power failures. When a big storm blows up and utility lines go down, your home and business are suddenly robbed of precious energy. Memories of the huge blackouts that struck the Northeast in 1965 and the New York City area in 1977 also call to mind the widespread looting of businesses made defenseless by the loss of electrical power. If that is one of your concerns, you will be glad to know that you can connect a portable ac generator to your home or business power wiring, to provide at least some of the essential energy needed until utility power is restored.

What Is Available

Portable ac generators are especially compact members of a larger family of gas-driven power plants. The distinguishing features of portables are their compactness, relatively low weight (ranging from about 60 to 130 pounds) and comparatively small engine ratings (about 3 to 7 horsepower). In this range, output electrical power ratings range from 750 watts for the smallest models to 3500 watts for the huskiest models. Some are compact enough to take with you in the trunk of a full-size car, to stow in a camper, or to keep handy in the garage or tool shed. These compact models can also serve as a back-up generator to furnish essential power if commercial power goes out. Larger models are available on wheeled carts, for ease in getting the generator into position at the connection point.

Important Features to Look For

There are a number of important features to look for in a power generator and these are outlined in the sections that follow.

Size and Weight. Design and construction play an important part in determining the shape and bulk of the end product. Compact generators require much smaller antivibration mounts than other types which reduces size. Physical size alone, however, should never be the single determinant of the generator you purchase.

Table 6-1. Power Requirements of Common Electrical Devices.

Appliance	Average Wattage Rating
Blender	390
Broiler	1450
Carving knife	100
Coffee maker	900
Deep fryer	1450
Dishwasher	1290
Egg cooker	500
Frying pan	1200
Hot plate	1250
Mixer	125
Oven, microwave	1450
Range with oven	12,200
Roaster	1300
Sandwich grill	1150
Toaster	1150
Trash compactor	400
Waffle iron	1100
Waste disposer	440
Freezer (15 cu. ft.)	340
Freezer (frostless—15 cu. ft.)	440
Refrigerator (12 cu. ft.)	340
Refrigerator (frostless, 12 cu. ft.)	320
Refrigerator/freezer (14 cu. ft.)	325
Refrigerator/freezer (frostless)	600
Clothes dryer	4850
Iron (hand)	1000
Washing machine (automatic)	500
Washing machine (non-automatic)	280
Water heater (standard)	2465
Water heater (quick recovery)	4475
Water pump	460
Air cleaner	50
Air conditioner (room)	1500
Bed covering	175
Dehumidifier	275
Fan (attic)	370
Fan (furnace)	290
Fan (windows)	200
Heater (portable)	1320
Heating pad	65
Humidifier	175
Oil burner or stoker	265
Radio	70
Radio/record player	100
Television (black and white, tube)	160
Television (black and white, solid)	55
Television (color, tube type)	300
Television (color, solid-state)	200
Clock	2
Floor polisher	300
Sewing machine	75

Vacuum cleaner	630
Electric brooder	100 (higher)
Milking machine	250 (higher)
Milk cooler	500 (higher)
Milk pump	200 (higher)
Barn fan	125 (higher)
Barn cleaner	1500 (higher)
Feed conveyor	375 (higher)
Elevator	375 (higher)

Power Output in Watts. This is the real key in buying the right generator and you should determine the need for the maximum power you will require well in advance of considering physical size. Table 6-1 will help. It lists the wattages consumed by various home and farm electrical devices. Look up the wattages of devices you intend to power. If more than one will be powered at a time, add their wattages together. Next, add a safety factor of 20 percent, to account for variations in load presented to the generator by different devices during start-up. The total wattage you sum up will be the wattage required. Buy the generator that has the nearest, *higher* wattage rating. That is, if you add up a need for 1600 watts output, do not try to get by with a 1500-watt generator. You will shorten its life and always be unhappy with portable power, because the normal load exceeds the generator's capacity. Rather, buy a 2000-watt generator. It will run smoother for longer periods, with fewer maintenance problems, because it has reserved capacity beyond the needed power.

Noise Level. The engine that furnishes drive power to the generator may be running outdoors for prolonged periods in some cases. Choose a unit with an effective muffler system that reduces engine noise to the lowest practical level. Consider where and when you will be using your generator. If it will be running in a location that is far from earshot of you or your neighbors, noise level is not important. But, if you will use it within 100 feet of your own location, or that of your neighbors, get the quietest design you can find.

Fuel Capacity. Too small a fuel capacity means annoying stop-and-go refueling periods. As a minimum, your generator should run for at least 90 minutes at rated power output on a single fueling.

Interference. You may wish to power radios, TV sets, or communications equipment from your generator. If so, be sure that the ignition system of its engine has the latest interference suppression devices, so that the generator does not become a source of electrical interference in the area of use.

Electrical Outlets and Extra Features. To simplify connecting the generator to the device it will power, built-in ac outlets are provided. These may be the same as standard duplex receptacles found in home wall outlets for those generators designed to give 120-volt, 60 Hz ac power. Some generators provide more than one outlet, and some offer both 120 and 240-volt output. But, remember that the total power delivered to all devices plugged into these outlets must not exceed the generator's output wattage rating. For short circuit protection, look for fuses or a circuit breaker on the unit. Also, if you need to charge batteries from time to time, you may want to consider a generator that provides a 12-Vdc output. Some generators can provide up to 8 amperes of charging power for batteries in outdoor equipment, RVs, and remote equipment. This is a real convenience where on-the-spot battery recharging is necessary in remote location.

USING A PORTABLE AC GENERATOR

A portable ac generator is a great convenience in powering outdoor tools such as electric trimmers, electric mowers, farm machines, and for the comfort of electrical home conveniences while camping. Setting up a portable generator is simple: choose a fairly level surface in the open air, fill the gas tank with clean, fresh, regular-grade gasoline, and be sure to check the engine crankcase for proper lubricant level. (Typically, SAE 30 is used in summer, while SAE 5W-20 or 5W-30, is the choice for winter operation.)

Starting

Some models have a fuel shut-off valve. This must be opened to allow fuel to reach the carburetor of the engine. (It is normally closed for safety, either when shutting off the generator engine, or when the generator is in transit, to prevent fuel spillage.) For easy starting, unplug any devices from the generator's outlets, or turn their power switches off. (It is very difficult to start a generator when a load is connected.) Next, close the choke (if cold-starting), and operate the generator's starting mechanism (typically, a pull-cord recoil starter). Once the engine starts, allow a slight warmup, then, gradually ease the choke to the open position.

Running

Once the engine has come up to speed, the generator will be

providing full output to the outlets. You may now plug in, or turn on, the electrical devices to be powered by the generator. If current drain is substantial, you may notice a variation in the engine sound at turn-on, as the generator commences to supply power to the load. This should smooth out in a few seconds as current flow stabilizes.

Refueling

If gasoline runs out, the generator engine will stop. Cautiously refuel, remembering that the engine is still warm and may cause the flammable fuel to volatize on contact. Restart without the choke, remembering to disconnect the plug of the device being powered, so as to not increase starting effort. There is danger in refueling a generator while it is running, and it is wisest to allow the fuel to run out and the generator to stop before adding fuel.

Generator Stop

Most portables include a quick-stop feature to permit rapid shut-off of the generator. This is usually a lever that momentarily shorts out the engine spark plug to ground, thus depriving the engine of ignition voltage and causing it to stop. The preferred method of stopping is to unplug or turn off all the load and allow the engine to run for a few minutes so as to reduce its operating temperature before engine shut-off.

Battery Recharging

If the portable you are using includes a battery-recharge feature, proceed as follows: connect the positive terminal of the battery to the positive terminal of the generator and the negative terminal of the battery to the negative terminal of the generator. Start the engine. The battery will receive an initial high charge of several amperes that slows to a trickle as the battery comes up to strength over a period of time.

MAINTAINING YOUR GENERATOR

The following are some general rules for maintaining your generator and the engine that drives it.

Lubrication

Most generators require periodic engine oil change to ensure good lubrication (usually after every 25 hours of operation). It will also be necessary to wash the carburetor air cleaner in gasoline to free it of airborne dirt and particles, and then to re-oil it. The manufacturer's manual will specify the exact lubrication requirement of your generator.

Spark Plug

About every 100 operating hours, you should remove the spark plug of your generator's engine and clean it with a wire brush to remove combustion deposits. If necessary, square up the electrodes and regap (typically, to 0.03 inches) with a feeler gauge. You can also replace the spark plug with a properly gapped new plug at 100-hour intervals, to be sure of efficient ignition and full power output.

Transport

If your generator provides a fuel shut-off valve and tight-fitting gas tank cap, you can transport your unit after taking all precautions to prevent fuel spillage. You could also siphon any residual fuel from the gas tank to prevent hazardous fuel spills.

Storage

Your generator should not be stored with fuel in the tank because gasoline goes "stale" and releases waxy residues over prolonged storage periods. The danger is that these residues will find their way into the carburetor, causing blockage and no-start problems, necessitating a messy clean-up operation. Before "mothballing" your generator, run the engine to use up any residual fuel; then, sponge up any leftovers by wrapping a cloth around a stick and using this to soak up any excess gasoline in the tank. Discard the cloth safely afterwards. To lubricate the engine properly during storage, remove the spark plug and pour into its hole one tablespoon of SAE 30 oil. Replace the plug and pull the starter several times to distribute an oil film over the engine cylinder and piston.

COMMERCIALLY AVAILABLE GENERATORS

In the selections that follow, several commercially available generators will be described. This will give you an idea of what you might want to buy for your home or camp.

Winco Dynamight Generator—750 Watts

This handy, lightweight generator generates a maximum of 900 watts of 60 Hz ac power to run television sets, appliances,

Fig. 6-1. Dynamight K900 generator provides 600 watts average power (courtesy Dyna Technology, Incorporated).

emergency lights, power tools, and furnace motors. It also provides 6 amperes of dc power to charge batteries, all at the same time.

The Dynamight K900 generator is shown in Fig. 6-1 with two 120-Vac outlets and the 12-Vdc output for charging auto batteries. The unit weighs only 44 pounds and can be easily carried to any location where there is a power outage or power is needed in a remote location. Figure 6-2 shows the K900 being carried easily with one hand. Homeowners, outdoor enthusiasts, boaters, and farmers will find the unit fills the need for temporary stand-by power.

The K900 can be shipped by UPS and it features a USDA Forestry Approved muffler for safe and super-quiet running. The K900 offers automatic reset overload protection, automatic throttle control and has a 76 cc 4-cycle Kawasaki engine to drive the generator.

Figure 6-3 shows the Dynamight being started with its recoil starter after it has been carried to an operating site where emergency power is needed. Figure 6-4 shows the unit providing a quick charge to an auto battery. Remember, a battery charging system such as this is intended to recharge weak batteries and not for starting vehicles having dead batteries. Charge the battery for

15 to 30 minutes before trying to start the car. Table 6-2 shows the specifications for the K900 system.

The Dynamight K900 is manufactured by Winco, Division of

Fig. 6-2. The K900 is light enough to carry with a single hand.

Fig. 6-3. The K900 generator starts easily with the one-pull starter.

Dyna Technology, Incorporated, 7850 Metro Parkway, Minneapolis, Minnesota 55420.

Clinton Model 500 750-Watt Generator

This small, very portable generator, Clinton Model 500-750W, produces a 750 watt output at 6.25 amperes continuous duty. Shown in Fig. 6-5, the 60 Hz unit has a single handle for carrying to a site. Because of its use of premixed oil and fuel, the 3-hp engine can be generated at any angle. The engine has only three internal moving parts which are lubricated on every piston stroke.

The Clinton unit weighs only 48 pounds and produces enough energy to power tools, television sets, appliances, lights, pumps, spot lights, electric heaters, and radios. Figure 6-6 shows the Clinton 500-750W supplying power to an electric saw with the cord plugged directly into the unit. The engine has a 1¼-quart fuel tank with a 2½-gallon tank available.

The generator has fused circuit protection and two ac 120-volt receptacles. The unit shown has an optional battery charging fused circuit and terminals. The unit standard features include an easy-pull automatic rewind starter, firm-grip handle located for best carrying balance, and four rubber feet to prevent vibration movement and damage to floor surfaces.

Fig. 6-4. The Dynamight K900 charging a car battery with its 12-Vdc output.

Table 6-2. K900 System Specifications.

Item	Description
Surge or maximum rating (ac)	900 watts
Continuous rating (ac)	750 watts, 115 volts @ 6.5 amperes
Battery charging dc	6 amperes at 12 Vdc
ac and dc at same time	Yes
Battery cables	Standard
USDA approved low-noise spark arrestor muffler	Yes
Number of ac outlets	Two
Overload protection	Automatic
Output indication light	LED
Engine make	Kawasaki
Engine capacity	76 cc, 4-cycle
Throttle control	Automatic
On-Off switch	Yes
Weight	44 pounds
Rated output	1.25 hp at 3600 rpm
Spark plug	NGK BM-6A

The Clinton Model 500-750W Generator is from Clinton Engines Corporation, Maquaketa, Iowa 52060.

AG-Tronic Powermate Electric Generator

The Powermate rotary field generator by AG-Tronic, Incorpo-

Fig. 6-5. The Clinton Generator Model 500-750W produces 750 watts of ac power (courtesy Clinton Engine Corporation).

Fig. 6-6. Clinton 750-watt generator supplying 115 Vac power to electric saw (courtesy Clinton Engine Corporation).

rated, is available in power sizes from 1250 to 5000 watts. The lightweight, low cost AG-Tronic Powermate is a portable power plant designed to supply you with electrical power anywhere, any time. The Powermate is unbeatable for applications such as emergency standby, power tool operation, and running repair or recreational equipment in the field.

The 3000-watt Model No. 433003 Powermate is shown in Fig.

Fig. 6-7. The Powermate by AG-Tronic, Incorporated, provides 3000 watts at 120/240-Vac (courtesy AG-Tronic, Incorporated).

6-7. This unit is designed for general home and construction use and features AG-Tronic's advanced rotary field generator and rugged Briggs and Stratton engine. The Powermates are built for continuous service at full rated capacity and incorporate many quality design and construction features. The 3000-watt generator offers sufficient home standby power to run furnace, refrigerator, freezer,

Table 6-3. Specifications for 3000-Watt Powermate.

Item	Description
Model No.	433003
Watts	3000
Volts (ac)	120/240
Amperes	25/13
Horsepower	7
Net weight	100 pounds
Shipping weight	107 pounds
Length	22 inches
Width	19 inches
Height	18 inches
Outlets	Duplex, 120 Vac, 20 amperes, Twistlock, 240 Vac, 3-prong receptable

Fig. 6-8. Sears 3000-watt long-run portable generator. Model shown has 5-gallon fuel tank.

sump pump, and some lights. It can also be used for light-to-medium duty construction work. Table 6-3 provides additional specifications on the 3000-watt Powermate. A pull cord with recoil is used to start the power unit. The unit is stopped by pressing the stop switch against the spark plug.

Powermate portable power generators are made by AG-Tronic, Incorporated, 364 Airport Road, Kearney, Nebraska 68847.

Sears Long-Run Portable Generator

This heavy-duty power plant provides attention-free electricity for extended periods of time. This unit features an auxiliary fuel and oil reserve for additional unattended running time. Shown in Fig. 6-8, the fuel system includes a fuel pump, selector valve, hoses, and a 5-gallon gas tank with a quick-disconnect feed hose.

This unit provides 3000 watts of 60 Hz at 120/240 Vac. The unit is equipped with an air-cooled, 4-cycle, 3600 rpm engine and has a recoil start. Figure 6-9 shows the alternator control panel and power outlet receptacles with a power cable being inserted in the 240-volt outlet. The unit also has two 120-volt parallel blade outputs and one 120-volt three-prong outlet.

Battery charging terminals on the control panel supply up to 8 amperes at 12 Vdc for recharging vehicle batteries. Cables and clips are provided. The 12 Vdc should be used for charging batteries, not for starting autos as the current output is not sufficient.

An optional wheel kit is available, which makes it easy to move the power unit from place to place. This kit is complete with wheels, handle extension, and necessary hardware for attaching to the power unit. A close up of the engine choke lever is shown in Fig.

130

6-10, as well as the valve to the fuel supply.

Specifications for the Sears 3000-watt generator are shown in Table 6-4. This power generator is available from most local Sears retail outlets.

CONNECTING YOUR GENERATOR
TO A HOUSEHOLD ELECTRICAL SYSTEM

The generators we have discussed in this chapter produced 60 Hz power from 750 to 3000 watts. While most of the time we might use a portable generator for outdoor activities such as camping, fishing, amateur radio field-day operations, hunting, building construction, and repair, we might also want to consider using the higher-powered generators to supply emergency power to our homes.

Main Components for Home Supply

If you want the ability to take over supplying your own electricity when utility power fails, you should plan to have the required

Fig. 6-9. Power cable being inserted into 240-Vac outlet of Sears Craftsman generator.

Fig. 6-10. Choke lever adjustment and supply valve for dual fuel supply for Craftsman generator.

transfer and connection devices installed right after you buy your generator. But remember this important rule: *Don't do it yourself!* Electrical codes require that a licensed electrician make the connections. The risk of electrical shock is great while making the installation; you also risk incurring the wrath of your local utility company, city building inspectors, and insurance company if something goes wrong. Be content to be the prudent planner and operator of your own home power station. Leave the installation to a licensed professional.

In Fig. 6-11, we see the main components that make up the power transfer capability required to provide your home with local utility power or your own home-generated power. These components are a connection box, transfer switch box, and a connection cord set to run to your generator outside your house.

Transfer Switch

The transfer switch is the most important device in the scheme

132

as it permits power to go to your home circuit wiring from either of two sources—the power company or your home generator—but never both of them at the same time. Feedback of your generator voltage into the power lines could endanger the life of anyone working on the lines and would probably destroy your generator through a short circuit. The transfer switch also prevents accidental re-energizing of your own equipment, which could destroy your standby generator when regular power is restored by the power company. For single-phase power as we discuss here, the transfer switch is a double-pole, double-throw type.

Connection Box and Cord Set

The connection box shown in Fig. 6-11 is a means of connecting your power generator into the transfer switch. The cord set connects the generator output to the connection box with suitable plugs on each end and would be connected at such time when emergency power is needed. The connection box is mounted outside your home, ready for you to connect your generator before you start the power unit and throw the transfer switch from utility power to home power.

Table 6-4. Specifications for Sears Portable Alternator.

Item	Description
Wattage	3000
ac voltage	120/240
Amperage	25 @ 120 volts
	12.5 @ 240 volts
Frequency	60 Hz
Phase	Single Phase
Battery charging	12 Vdc, max at 8 ampere, taper to trickle charge
Voltage regulation	Solid-state maintaining 120 Vac ($\pm$ 2 percent) at 60 Hz
Electrical outlets	Two 120 volts, parallel blade, rated at 15 amperes each
	One 120 volts 3-prong, rated at 30 amperes
	One 240 volts 4-prong, rated at 20 amperes
Engine	Craftsman HM70
Horsepower	7
Governor speed	3600 rpm
Crankcase oil capacity	11½ pints
Fuel tank capacity	4 quarts
Spark plug	Champion RJ-17LM
Starting	Manual, recoil rope

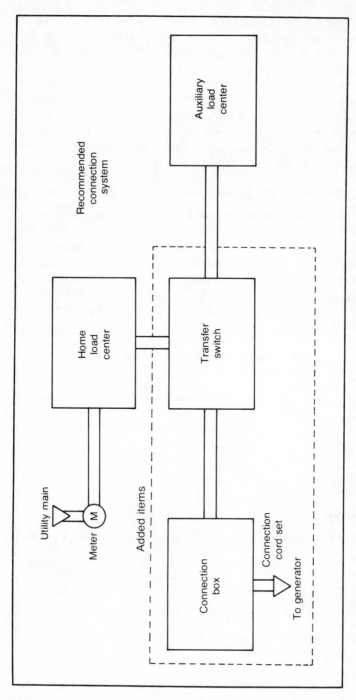

Fig. 6-11. Block diagram of recommended system for connecting home power generator into home wiring.

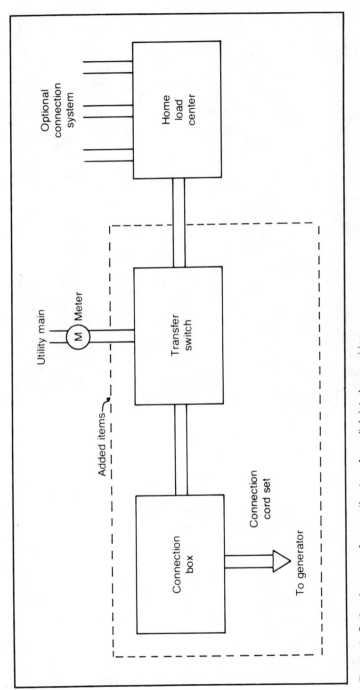

Fig. 6-12. Optional means of connecting transfer switch into home wiring.

Fig. 6-13. Sears 100-ampere transfer switch. Connection openings to home load are shown.

An alternate means of connecting the transfer switch is shown in Fig. 6-12. This is the simpler of the installations. It does not use the auxiliary load center to the home, which requires separate wire connections to the devices to be powered.

Sears 100 Ampere Manual Transfer Switch with Box and Cord Set

The Sears Transfer Switch is shown in Fig. 6-13 and is equipped with a 100-ampere circuit breaker when connected to utility power and a 60-ampere circuit breaker when connected to the power generator. The switch is mechanically interlocked so that only one of the circuits can be in use at a time. The schematic diagram for the 100-ampere transfer switch is shown in Fig. 6-14. Note how the 100-ampere and the 60-ampere breakers are interlocked, so that one is open and the other closed.

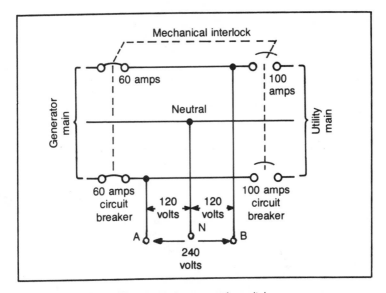

Fig. 6-14. Schematic diagram for Sears transfer switch.

A 100-ampere transfer switch should be used when the commercial input to your home is less than 100 amperes. If the commercial amperage input is greater than 100 amperes, but less than 200 amperes, use a 200-ampere transfer switch. When the transfer switch is switched to the emergency position position (with your

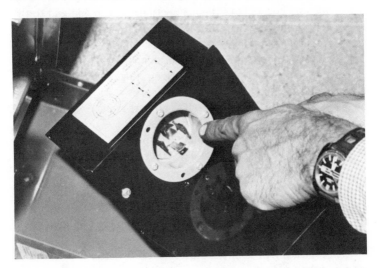

Fig. 6-15. Sears connection box for the power transfer system is shown with four-prong Twistlock Flanged Inlet.

137

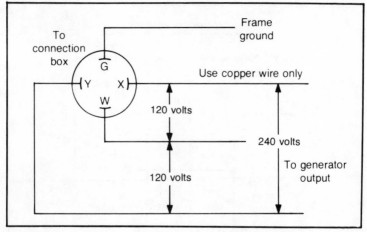

Fig. 6-16. Wiring diagram for Sears Cord Set connecting the connection box and the generator.

generator running), the generator's ac power output will pass through a 60-ampere circuit breaker.

The connection box of the power transfer system is shown in Fig. 6-15. This box is waterproof, as are the other system boxes, and it contains the four-prong Twistlock Flanged Inlet 120/240-volt input to the system.

The wiring diagram for the connection cord is shown in Fig. 6-16. One end connects to the connection box and one end to the generator output 240-volt three-wire and ground receptacle. The cord set supplied is a four-wire, 240-volt cord, rated at 20 amperes. It is to be plugged into the 240-volt outlet on the generator's control panel. It is important that the amperage rating of your cord set not be exceeded. The cord is rated at 20 amperes at 240 volts. This can be given in watts of power, as well as amperes, since watts is equal to volts times amperes. As an example:

$$\text{watts} = \text{volts} \times \text{amperes}$$
$$\text{watts} = 240 \times 20$$
$$\text{watts} = 4800$$

As shown above, you should never run tools or appliances requiring more than 4800 watts of power. You never want to exceed the output rating of your generator, so add up the wattages of all tools and equipment to be used at one time. This total should not exceed the rating of your generator or 4800 watts, the maximum rating of the cord set.

Figure 6-17 shows the local home load center power box

hardware with connections that feed the transfer switch. One channel is for input to the transfer switch and the other is for the output from the transfer switch. Refer to Fig. 6-13 to see both of them connected together.

Operation of the Transfer Switch

When the utility company is feeding power, the transfer switch routes it to the primary home load center for distribution to all

Fig. 6-17. Mechanical hardware connection between home load center and transfer switch.

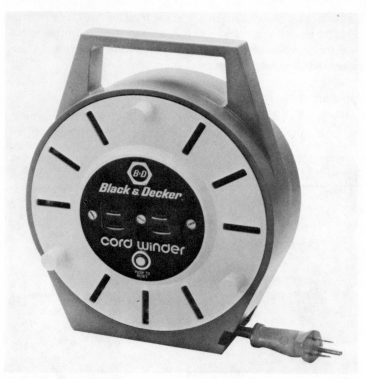

Fig. 6-18. Black and Decker 25-foot Cord Winder with circuit breaker (courtesy Black and Decker).

circuits in the household. However, should the commercial power fail, wheel the ac generator to the outside connection box, connect it with the cord set, and start the engine. After a few minutes when the engine has stabilized, flip the transfer switch to its opposite position. This now connects the home generator to the essential circuits in your home. Remember, all the circuits in your home are completely disconnected from the utility company main by the transfer switch so that there will be no problem when the utility power comes back on.

Black and Decker Cord Winder

The Cord Winder permits you to feed power to various appliances up to 25 feet away from your generator or any other electrical outlet inside or outside your home, garage, marina, or trailer camp. Shown in Fig. 6-18, the unit is provided with two 120-Vac outlets so that you can provide power to a light and one other device such as a drill or saw for use at night.

The 25-foot extension cord is coiled inside a lightweight, crack-resistant, housing. The cord can be pulled out to any length up to 25 feet and the housing protects the cord and eliminates tangling. The heavy-duty 14 gauge cord stays flexible in cold weather and has a built-in 10-ampere circuit breaker for the two electrical outlets. The breaker can be reset from the panel. The handle makes rewinding the cord easy and the unit is compact, easy to carry and store. The unit is UL-listed.

The Cord Winder is manufactured by Black and Decker, Consumer Assessory Division, Hampstead, Maryland 21074 and is available from local suppliers.

Black and Decker 50-Foot Extension Cord

The 50-foot extension cord is shown in Fig. 6-19 and is a 3-wire 16 gauge extension cord for use with any ac outlet or power generator. The vinyl insulation is orange-colored and remains flexi-

Fig. 6-19. Black and Decker 50-foot three-conductor extension cord (courtesy Black and Decker).

ble in cold weather. The 50-foot extension cord is Model No. 88-050 and is available from local outlet hardware stores.

AUXILIARY GENERATOR SAFETY CHECKLIST

Some generator safety considerations are listed below. These should be kept in mind when selecting a generator and a location for its operation.

- Never operate the generator in an enclosed space. Provide adequate ventilation for generator equipment cooling, and ventilation for removal of exhaust fumes.
- Locate generating equipment away from combustible materials. Provide adequate space for maintenance access if installed in permanent location.
- Ensure that engine fumes will not enter premises through ventilation intakes or windows.
- Provide muffling and vibration isolation for exhaust from engine system.
- Remember that alternators produce lethal voltages; they should be treated with respect.
- Prevent backfeed of utility company power into generator system.
- Provide guards on electrical equipment such as fans.
- Provide antifreeze protection for a water-cooled engine.
- Ensure sufficient fuel for 90 minutes operation at full power load.
- Take precaution to ensure an internally mounted fuel tank will not have to be refueled during unit operation.
- Never fill the gas tank while the engine is running. Gasoline spillage on a hot engine can cause a fire or an explosion.
- Store fuel in accordance with all applicable fire safety codes and insurer's recommendations.
- Perform preventative maintenance regularly.
- Check engine oil level each time you fill the gas tank.
- Ensure that all wiring and system components are kept dry and free of dust and other contaminants.
- Maintain power cords in good condition.
- Perform periodic operational tests under load.
- Test operation of the transfer and start up equipment regularly.
- Never attempt to change engine speed without proper knowledge and equipment. Incorrect engine rpm is not only

dangerous, but can damage the alternator or the equipment it powers.

- Emphasize that all nonessential loads must be shut off, such as heaters, hot plates, coffee pots.
- Study the contents of the manufacturer's manual carefully before operating your generator.

7 Batteries, Chargers, and Tester Accessories

The battery is probably the first device invented that started the modern revolution known as electronics—considering that the electric eel was not invented by man nor was lightning. Batteries touch our everyday lives in familiar ways such as starting our autos, powering our transistor radios and portable cassette tape players, watches, calculators, smoke detectors, hearing aids, and the like. They also affect our lives in ways that we are not directly aware of such as powering the world's great telephone systems. The telephone never goes out when there is a power failure because it was designed around the use of batteries over a hundred years ago by Alexander G. Bell.

The battery was invented by Count Alessandro Volta (1745-1827) who was born in Como, Italy. His discovery of the decomposition of water by an electrical current laid the foundation of electrochemistry. In his honor, the volt, a unit of electrical measurement, was named for him. Along with pioneering work in what was to later become electronics, he invented the electrical condenser, or capacitor, which also plays an essential role in electronics.

BATTERIES

Strictly speaking, the familiar flashlight "battery" is not really a battery but a *cell*. Figure 7-1 shows the familiar flashlight batteries which put out 1½ Vdc, as do most dry cells used in flashlights. The two-cell or three-cell flashlights have two or three 1½-volt cells

Fig. 7-1. Building block of a battery: a single cell. On right, 1½-volt D cell; left, 1½ volt penlight cell.

that supply voltage to the bulb of 3 or 4½ volts. A *battery* is any combination of two or more *cells*, like the one in your automobile, in which six 2-volt cells are connected in series to form a 12-volt battery.

Figure 7-2 shows a 6-volt lantern battery which is used to power larger flashlights to produce a more brilliant light. This battery has four 1½-volt cells connected in series to produce 6 Vdc. Figure 7-3 shows the schematic diagram of a cell and a battery. The basic cell is a simple electrochemical device having three essential elements, as shown in Fig. 7-4: an *anode*, a *cathode*, and an *electrolyte*. All other components are simply to hold things in place, protect the assembly, and perhaps seal in the electrolyte.

A chemical reaction between the cathode and the electrolyte causes the cathode to build up a huge excess of electrons. Another reaction, of different nature occurs at the same time between the anode and the electrolyte. The net effect of the two reactions is to develop a voltage difference between the anode and cathode that can range from as low as 0.8 volts to as high as 2.4 volts, depending on the materials used in the construction of the anode and cathode, and type of electrolyte. This voltage can then be used to drive current through an external circuit—that is, to pull electrons from the cathode (negative terminal) through the external load to the anode (positive terminal).

All cells work on this basic electrochemical process, regardless of their type. There are two kinds of cells or batteries to which

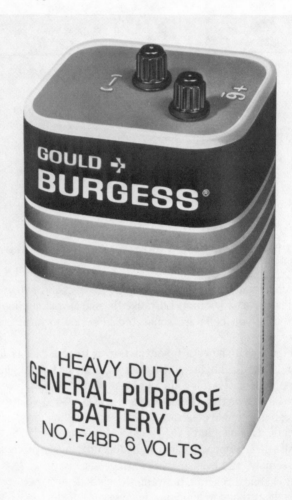

Fig. 7-2. Size F 6-volt lantern battery. Consists of four 1½-volt cells connected in series.

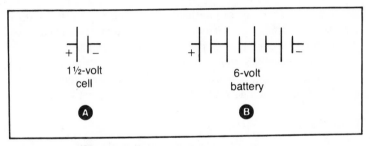

Fig. 7-3. Schematic for cell (a), battery (b).

all electrochemical sources of voltage belong: primary and secondary. In the primary cell or battery, energy is obtained from the electrochemical reaction, and when withdrawn from the source, it can never be replaced. Such a cell or battery may be "revitalized" by recharging but the effect is temporary, inefficient and undependable. The primary cell is used until its voltage output begins to drop to an unuseable level and it is then thrown away. Examples of the primary type are flashlight cells, quartz watch cells, transistor radio batteries, and calculator batteries.

In the secondary cell or battery, energy is stored by electrochemical action when the cell or battery is "charged," by forcing

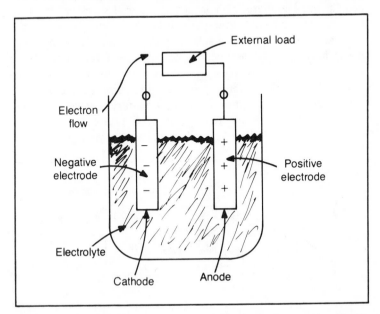

Fig. 7-4. Basic cell structure. Electrons flow (move) from minus to plus through load.

current through it (from another energy source) in the reverse direction to normal load current. Many years ago, storage batteries were called "accumulators" because they accumulated and stored electrical energy. Examples of the secondary cell or battery are automobile, marine, and aircraft batteries, rechargeable flashlight cells, and rechargeable transistor and calculator batteries.

CHARGERS

A *charger* is an electrical device that is used to put energy into a cell. It is a simple ac to dc power supply in which the normal output voltage is slightly higher than that of a fully charged battery. The current, however, is limited so that it will not become excessive if the charger is connected to a "dead" (fully discharged) battery. The charger usually consists of three elements: a *transformer* to step down the ac voltage from the power line to the desired voltage; a *rectifier* to change ac voltage to dc voltage, and a *current-limiting circuit* to prevent excessive charging current to flow into the cell or battery.

Figure 7-5 shows the circuit diagram for a basic battery charger. The transformer not only steps the voltage down to a lower level but it also electrically isolates the charging circuit from the main ac lines so there is no danger of personal shock or electrical damage. The current-limiting circuit can be as simple as a single resistor or as complex as a self-programming multitransistor regulator having almost as many components as a regulated power supply. The more complex circuits are used to shorten the recharging time by maintaining the charging current at or near the maximum allowable safe limit until the battery is very nearly recharged. When a simple resistor is used as a current limiting device, the current falls off steadily as the battery voltage builds up.

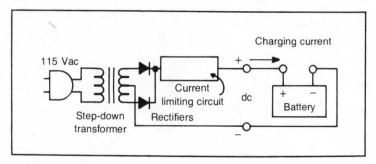

Fig. 7-5. Three basic elements of a battery charger: transformer, rectifier, current-limiting circuit.

The charge rate thus continually tapers off which doubles or triples the charging time.

TYPES OF PRIMARY BATTERIES

There are a number of different types of disposable primary batteries and all of them are of the so-called "dry" type. That is, the electrolyte is not a liquid, but a paste, gel, or other semisolid. There are many different combinations of anode materials, cathode materials, and electrolyte in use.

The ordinary flashlight cell, the carbon-zinc cell, has a carbon anode (actually, graphite and filler), a zinc cathode (actually, zinc-plated steel), and a witches' brew of various acid salts, oxides, stabilizers, and water for the electrolyte.

Alkaline batteries are dry cells. They provide longer life with higher cost.

Mercury batteries are also alkaline, but with even longer life and an even higher cost. They are used for watches and pocket calculators.

Lithium batteries have the longest life and also the highest cost of all the cells. They have a shelf life in excess of seven years and are now used almost exclusively in implanted heart pace-makers.

There are other systems, too, designed to optimize peak-current capability and high-temperature operation.

TYPES OF SECONDARY BATTERIES

Rechargeable batteries fall into several categories. We will discuss the most popular types.

Fluid Electrolyte

This is a fancy name for the battery used in your car, which is of the lead-acid type. It consists of six cells of 2 volts each to provide 12-volts for your auto's electrical system. They can be recharged for a number of years.

Nickel-Cadmium

The NiCad uses a gel electrolyte and is widely used in aircraft electronics because it will not spill, and can be sealed permanently. The most common of the rechargeable gel electrolyte types, it is widely used in small instruments, calculators, and portable radios. It is also available in a bewildering array of ampere-hour, voltage, and peak-current ratings, estimated number of charge/discharge cycles, physical sizes, and environmental ratings.

BATTERY PROPERTIES

There are a number of properties inherent in batteries that determine the type of service you will get from them. At certain times, these are very crucial as discussed below.

Capacity

The capacity of a battery, expressed in ampere-hours (AH), is the total amount of electrical energy available from a fully charged cell. Its value depends on the discharge current, the temperature during discharge, the final cutoff voltage, and the general history of the battery. When a battery discharges current at a constant rate, such as starting your auto engine in cold weather, its capacity changes according to the amperage load. The capacity of a battery increases (stays high) when the discharge current is less than a certain AH rate and decreases when the discharge current is higher.

Discharge

During discharge, the voltage will decrease. The more current drawn out of the battery, the faster the voltage will drop. If your car will not start immediately, do not just continue to crank it; wait 10 to 20 seconds and then try again. You will rest your battery and also let your starter cool off. Remember, the starter is for intermittent starts only, and is not to be run as a motor.

Open Circuit Voltage

Open circuit voltage varies according to ambient temperature and remaining capacity. Generally, it is determined by the specific gravity of the electrolyte. Discharge lowers the specific gravity; consequently, it is possible to determine the approximate remaining capacity of the battery from the open circuit voltage. Cars equipped with voltmeters or LED indicators are very handy in being able to tell the condition of their car battery. You can tell if the battery is gradually becoming weak.

Temperature

Capacity is a function of ambient temperature and rate of discharge. Most batteries have a rated capacity of 100 percent at 68° F. The capacity increases above this temperature and decreases as the temperature falls. The higher rate of discharge, the lower the available capacity. That is why cold winter car starts are so difficult on a battery, while summer starts are easy. Not only does the battery have less get up and go in winter but your engine has cold jelly for oil in the winter.

Shelf Life and Storage

If your battery has a low self-discharge rate, it will have a long shelf life. For most lead-acid batteries, this can be many months, while for dry batteries, it could be several years. The rate of self discharge varies with the ambient temperature. If you are not going to use dry cells for many months, store them in your refrigerator. When you take them out, they will be almost like new!

Life-Cycle Use

The number of charge/discharge cycles depends on the capacity taken from the battery (a function of discharge rate and depth of discharge, operating temperature, and the charging method). At nominal room temperature, a 30 percent depth of discharge can give you approximately 1200 charge/discharge cycles, while a discharge depth of 100 percent will give you about 180 cycles. If your car will not start easily and requires regular and continual cranking, you could damage your battery when all you might need is a spark plug change.

Life Standby Use

The float service life, or life expectancy under continuous charge, depends on the frequency and depth of discharge, the charge voltage, and the ambient temperature. A battery kept fully charged at room temperature should provide many years of service before its capacity drops compared to its original rating.

THE LEAD-ACID CELL

A battery produces an electric current through an external load by electrochemical reaction within the battery when the load is placed across the terminals of the battery. The cell or battery produces electricity when the battery discharges.

Discharge

We will look at the discharge cycle of a wet primary cell to see how it produces an electric current. We will assume the cell is fully charged as we see on the left in Fig. 7-6. At this time, the switch is open and we see that the electrolyte of sulfuric acid has a maximum

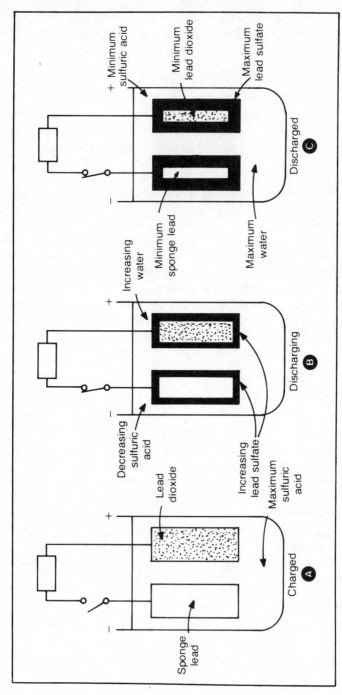

Fig. 7-6. How a cell discharges: (a) fully charged 2-volt cell with no load applied; (b) load placed accross charged cell causes cell discharge; (c) cell discharge with minimum sulfuric acid and maximum water.

amount of acid and a minimum amount of water. The negative plate has a maximum amount of sponge lead and the positive plate a maximum amount of lead dioxide.

When we close the circuit switch, in the middle of Fig. 7-6, the load is placed across the cell and current starts to flow through the load while chemical reaction takes place in the battery. This condition is the same as when you crank your car to start it or turn on the headlights with the engine not running. Several things start to happen as the cell discharges. Note that the amount of sulfuric acid starts to decrease while the amount of water starts to increase. The negative plate has decreasing sponge lead, while the positive plate has decreasing lead dioxide. Note that both negative and positive plates have increasing lead sulfate.

When the cell is discharged, like when you left your headlights on all day, the right-hand sketch shows the chemical condition within the battery. The negative plate will have minimum sponge lead (it started with maximum), while the positive plate will have minimum lead dioxide (it started with maximum). The electrolyte will have a minimum amount of sulfuric acid (it started with maximum) but it will have a maximum amount of water (it started with minimum). But note especially that both the negative and positive plates will have a maximum amount of lead sulfate. At this time, the cell will be discharged and produce little current. Maybe not even enough to honk your car horn or play the radio! So what do we do now?

Charge!

Let's take a look at what happens to a battery when we get a "jump" from another auto or put a charger across the run-down cell. To recharge the cell, all we have to do is simply connect a power source of higher voltage than the cell and reverse the direction of the current flow. This is what your alternator does in your car—it forces the current in a reverse direction through your battery and charges it up.

The chemical reaction that takes place is completely reverseable and the recharging may take from 1½ to 24 hours, depending on the depth of discharge and the charging voltage used. Let's look inside the discharged cell (battery) and see what happens. In Fig. 7-7 on the left, the switch is open and we have maximum water in the electrolyte and minimum sulfuric acid. Both plates have a maximum amount of lead sulfate and the battery is just plain flat!

In the center sketch, we close the switch and begin to charge

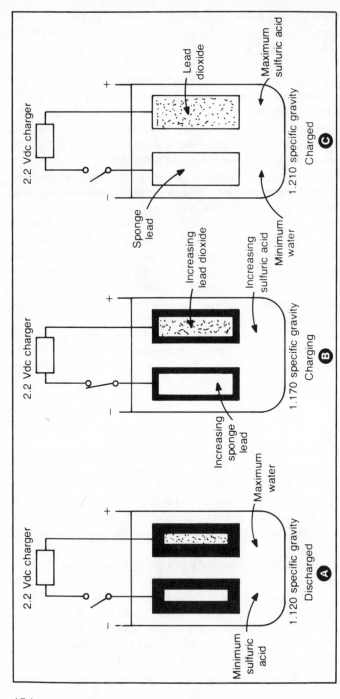

Fig. 7-7. How a cell gets recharged: (a) cell completely discharged; (b) specific gravity increases as cell gets charged; (c) specific gravity at 1.210 when cell is fully charged.

the cell. Note that the plates have increasing sponge lead and lead dioxide, respectively. Both plates also have decreasing lead sulfate, the water is decreasing, and the sulfuric is increasing. Finally, on the right, the cell is fully charged and we arrive back at the starting condition where we had maximum sulfuric acid, minimum water, the negative plate was sponge lead and the positive plate was lead dioxide. In a car battery situation, we go through this cargo-discharge cycle many thousands of times during the life time of each battery until finally it wears and will no longer hold a charge. Then it is time to replace it. You will notice the importance of this charge-discharge cycle when you drive at night with the headlights on and the car will not start the next morning. The headlight drain could not be replaced in the battery by the generator and it is weak the next morning when you give it the shock test with your starter!

While all this discharge-charge is going on, note what happens to the specific gravity of the electrolyte as it goes from minimum water to maximum water and back again. It goes from a high of 1.210 when charged to 1.120 when discharged and back up again when fully charged. That is why your garage man will measure the specific gravity of each cell to tell which one is going dead. This is really the acid test!

NICAD BATTERY AND CELL CONSIDERATIONS

Please note the following cautionary measures and charging tips regarding NiCads.

Caution

When using NiCad cells or batteries, there are some cautions that should be observed:

- Do not solder lead wires directly to the battery. Batteries with solder taps are available for your use.
- Do not short-circuit the battery. Excessive current will flow and damage the battery or possibly start a fire, as the short is heated.
- Do not dispose of battery in fire as it may explode due to gases developed.
- Do not disassemble the cell.
- Do not combine old cells with new ones. They will have different charge and discharge rates.
- Use cells of the same type so they will have the same charge and discharge rates.

Charging

The following conditions apply to charging NiCads.

Cyclic Use. Semiconstant or constant current charging at the 0.1-C (or C/10) rate for 15 hours is recommended. This means you should put the current back in the battery or cell at a slower rate than you took it out. Overcharging at C/10 for longer than 15 hours can be done at room temperature without causing damage to the cells.

Quick Charge. Cells ranging in size from small 1/3 AA to SC can also be quick-charged for four-and-a-half to six hours at the 0.25-C (C/3 to C/4) rate. Quick charging larger cells (C-cell and larger) require a controlled charge circuit because of the heat and gas generated during over charge.

Standby Use. A trickle charge of between 0.02 C and 0.05 C (C/50 to C/20) is sufficient to keep a battery fully charged. Within the temperature range of 32 to 113° F this charge rate will minimize heating effects during overcharge and prolong battery life.

BATTERIES FOR GENERAL USE

In the sections that follow, we will discuss various types of batteries available for general use in flashlights, lanterns, flashers, cameras, and standby emergency lights.

Polaroid Polapulse Planar Batteries

This wafer-thin 6-volt Pulapulse P100 battery serves as a power source for tools, instruments, electronic games, safety flashers, and numerous other devices. Figure 7-8 shows the Polapulse P100 battery as well as the recently introduced P500 Lithium battery. The two batteries are pictured alongside a regular size playing card. The lithium battery is even called the P500 Powercard because of its shape and size.

The P100 Polapulse battery weighs less than one ounce and measures only 3.8 × 3 inches and is only 0.18 inches thick. The 6-volt carbon-zinc power source incorporates four unitized 1.5-volt cells with only two contacts located on the same side of the battery to simplify device design and battery insertion. The planar design and simple contacts of the Polapulse battery virtually eliminate accidental reversal of polarity when the battery is inserted into a device. This had enabled original equipment manufacturers to save on materials, molding, packaging, and shipping costs.

To give you an idea of the success of this type battery, the Polapulse is an advanced version of the power source used in

Fig. 7-8. Polaroid Polapulse 6-volt battery on left and Lithium Powercard in center, compared to playing card on right (courtesy Polaroid Corporation).

Polaroid integral color films. Over 500 million packs of film, each containing the battery have been marketed throughout the world by Polaroid Corporation.

The Polaroid P500 Lithium Polarcard, pictured in the middle of Fig. 7-8, has a capacity comparable to four conventional AA alkaline batteries for most applications. It provides excellent operating time in a wide range of consumer electronic devices such as flashers, games, calculators, and compact radios.

Previously, lithium technology was available only in small 3-volt coin or button cells. The P500 develops 6 volts by means of two unitized 3-volt lithium cells, so that the leak-resistant, hermetically sealed Polaroid Powercard is the highest energy lithium battery available for replacement use. At a typical drain rate of many consumer devices of 20 mA, the battery is estimated to have a continuous operational life of more than 70 hours with a flat discharge curve. The Powercard's long life is attributed to the exceptionally high energy densities of lithium chemistry.

The Polarpulse and Polacard batteries are available at local retail stores.

Brinkmann Q-Beam Power Pack

The Brinkmann Power Pack discussed in Chapter 5 is a compact, highly portable source of rechargeable dc power. It is ideal for powering TV sets, CB radios, power tools, emergency lighting systems, and the like. It provides 12-Vdc output at current levels of 9 or 12 amperes. A built-in receptacle is provided for operating Q-Beam spotlights and other dc appliances equipped with a cigarette lighter plug. Figure 5-6 shows the power pack with its 120-Vac charger with cigarette plug for recharging the unit.

The Q-Beam Power Pack is manufactured by the Brinkmann Corporation, 4215 McEwen Road, Dallas, Texas 75234.

Yuasa Sealed Lead-Acid Rechargeable Battery

The Yuasa rechargeable batteries are available in a number of sizes with voltage output of 6 or 12 Vdc. These compact and efficient power sources are especially designed for long life, rugged durability, and dependable service. The batteries, because of the unique suspended electrolyte and venting system, can be mounted in any position without leakage or loss of capacity. Because of their sealed construction, the batteries can be operated, charged, and stored in any position.

In Fig. 7-9, we see 11 different types of batteries, ranging in size from the NP 1.2-6 that puts out 6 Vdc at 1.2 AH to the larger NP 24-12 that has a capacity of 12 Vdc at 24 AH. These batteries have a life expectancy of four to five years and a cycle use life of 250 to 1200 cycles.

Yuasa lead-acid rechargeable batteries are available from

Fig. 7-9. Yuasa lead-acid rechargeable batteries of various sizes (courtesy Yuasa America Incorporated).

Yuasa Battery (America), Incorporated, 9728 Alburtis Avenue, Sante Fe Springs, California 90670.

Kodak Disc Camera Lithium Battery

The lithium battery used in the Kodak Disc Camera is the result of intensive effort to perfect the battery that is, indeed, unique.

Search for a Battery. The battery was especially developed through a major research effort on the part of the Kodak and Matsushita Electric Company of Japan. The battery developed and operating in the camera powers the following activities. At the touch of a button, the disc camera will analyze the scene, set the proper exposure, activate the built-in flash, take the picture, advance the film, and recharge the flash, all within 1⅓ seconds. And to top it all, the camera is warranted for five years!

One of the interesting and crucial design decisions was the selection of a suitable "lifetime" power source. Most conventional cameras take a long time for their flash circuits to recharge. By the time the camera is ready, the picture or scene may be gone. Conventional camera units usually suffer from the dead-battery syndrome which usually means poor pictures or none at all.

The ideal power source for the Disc Camera had to provide fast recharging and extended life without the loss of capacity. Lithium-based batteries seemed to be the route to go; they are safe because they are unpressurized. And the chemical reaction that takes place is harmless. As lithium reacts with polycarbon-monofluoride, the original cell material is converted into carbon and lithium fluoride. Both materials are safe.

The Drop Test. The cells for the Disc Camera had to withstand a drop test from a height of 40 inches onto a concrete floor. After being dropped, the cells could sustain no internal shorts. The present battery has an internal electrode configuration so that the batteries successfully withstand the drop test.

The Short Test. A battery with low internal impedance is highly desireable, but this can cause problems. If a short circuit develops (that is, if the battery terminals are short-circuited externally), the battery can develop very high currents (on the order of 5 to 7 amperes). Such a high current can cause a high degree of heating which could cause some units to burst open. Such a hazard could not be presented to the consumer.

In the battery finally selected, the BR2.3A, a special separation material was found. The separator will reduce ion flow, and therefore, battery current when the battery heats up. That is, the modified separator shuts down the battery as soon as the internal battery temperature exceeds a predetermined limit.

An additional safety measure is also built into each camera power supply. The two batteries are connected in series by a thermal fuse to provide 6 volts. In this manner, the power source can be disabled by two means—one internal to the battery, and one external to the battery.

Lifetime of a Camera. Testing and data analysis by Kodak shows that the BR2.3A cells have a minimum 10-year shelf life. Their data also shows that the number of pictures taken by an average person during five years (the disc camera's warranty period) is considerably less that the capability of the camera.

Conclusion. The five disc cameras available from Kodak are Models 2000, 3000, 4000, 6000, and 8000. All the models use lithium batteries. In addition to the built-in strobe and motorized film advance mechanism, the camera also has some sophisticated electronics powered by the battery pack. Two integrated circuit (IC) chips determine when a flash is needed, as the electronic strobe will always flash if the light level falls below 125 foot-lamberts. The IC chips also accomplish all the required decision-making logic which includes timing and control for charging and discharging the flash capacitor, selecting the lens aperture and exposure speed, as well as advancing the film (rotating the disc). One of the IC chips includes the drive circuitry for the miniature motor with an average power of 2 watts and a peak current of 1 ampere.

The lithium battery that powers the Kodak Disc Camera is, indeed, a unique battery for a unique camera.

DIFFERENT SIZES OF BATTERIES

In the following portion of the chapter, we will discuss the different sizes of cells and batteries that are generally available. Figure 7-10 shows over two dozen batteries and cells made by different manufacturers. They are made in many sizes, shapes, voltages, and current output for various applications. Also shown are several battery testers and a battery or cell charger. There is bound to be a product to fit your requirement!

Type AA Cell

Rechargeable NiCad AA cells put out 1.2 volts, while carbon-zinc and alkaline AAs put out 1.5 volts. The NiCad 1.2-volt cells are on the far right in the back row in Fig. 7-11. Even though they supply

Fig. 7-10. A full line up of different types, sizes, shapes, and applications of various cells and batteries.

Fig. 7-11. A sampling of various rechargeable NiCad batteries. Shown are sizes AA, C, D, large button, stacked button, heavy-duty sealed high rite, and six units wired in series to provide 7.2 volts.

Fig. 7-12. A collection of some AA 1.5-volt cells including zinc-carbon and alkaline.

less voltage, they do put out more current, so toys calling for AA cells will run longer but slower on 1.2-volt cells.

Carbon-zinc and alkaline 1.5-volt cells are shown in Fig. 7-12. Pocket flashlights, safety devices for joggers, and many toys are designed to run off the 3.0 volts two of these cells will provide.

Type C Cell

The type C cell is next in size to the AA and is used in medium-sized flashlights, small toys, and medium-sized transistor radios. A rechargeable NiCad C cell puts out 1.2 volts and is shown third from the end on the right in Fig. 7-11. Carbon-zinc and alkaline C cells put out a nominal 1.5 volts. Figure 7-13 shows 1.5-volt C cells manufactured by many different battery companies.

Type D Cell

The D cell is the largest of the round, tubular type cells. It provides 1.2 volts in the NiCad rechargeable cell and 1.5 volts in the carbon-zinc and alkaline versions. The type D cell is the main stay of hand-held flashlights, larger electronic toys and is the most popular of all batteries. A D cell provides about .45 AH of energy. A rechargeable type D cell is shown second from the end on the left in Fig. 7-11.

Fig. 7-13. A group of representative 1.5-volt carbon-zinc and alkaline C cells.

Type F Lantern Battery

The lantern battery provides 6 volts to power lanterns and other devices that need 6 Vdc with large current capacity. Shown in Fig. 7-14, the type F battery is made up of four 1.5-volt cells connected in series. This type battery is available with spring terminals as shown or with screw and nut terminals. This type battery is available in a heavy-duty version which provides larger amounts of power for longer periods of time.

12-Volt Lantern Battery

This is perhaps the largest battery available before going to a

Fig. 7-14. Type F lantern battery provides 6 volts with spring terminals as shown or screw terminals.

wet cell or rechargeable gel-type battery. The 12-volt lantern battery is really two 6-volt F lantern batteries connected in series to provide 12 volts for heavy-duty work. The battery pictured in Fig. 7-15 has screw-type terminals and a heavy-duty flashlight can be mounted directly to the terminals as discussed earlier in Chapter 5.

GE Rechargeable Batteries

Various sizes of rechargeable nickel cadmium (NiCad) batteries manufactured by General Electric are shown in Fig. 7-16. These batteries have solder lug tab terminals and are ordinarily factory-installed in appliances such as electric shavers, cutting knives, flashlights, lanterns, radios, and toothbrushes. Note in Fig. 7-16 that only one of them, a D cell, has regular terminals. All of these batteries would provide moderate to heavy-duty service.

Special Purpose Batteries

Batteries suitable for special purpose and application are

Fig. 7-15. Heavy-duty 12-volt lantern battery with screw terminals (courtesy Gould).

Fig. 7-16. Assortment of rechargeable NiCad batteries by G.E. (courtesy General Electric Co.).

shown in Fig. 7-17. They range in voltage output from 3.0 volts to 22.5 volts and various configurations. The batteries shown are zinc-carbon or alkaline with snap-on or pressure terminals.

IC Chip Battery

Lithium batteries are used in implanted heart pacemakers because they can supply the necessary power for up to seven years. Lithium cells have been used recently with integrated circuit (IC) chips to provide standby power to retain memory on the chip. Rayovac, Incorporated, the battery supplier, has made the units reliable and long-lived so that Mostek, the IC chip manufacturer, can install them directly in the IC chip package. The stand by current for the chip is only 9 billionths of an ampere, and because of the low leakage current, the battery has a life time of about 2000 years, operating the chip without trickle charging!

TYPES OF CHARGERS

As discussed earlier, chargers are essential for replacing the energy taken out of rechargeable batteries or cells. The basic circuit diagram introduced in Fig. 7-5 works in a number of applications discussed below.

NiCad Battery Recharger

Figure 7-18 shows a NiCad battery recharger with its 115-Vac line cord. The cells and batteries in the figure can all be charged in this charger. Figure 7-19 shows the charger recharging two AA cells, two C cells and two D cells, all at the same time. Charging will not begin until the protective cover is closed to complete the circuit.

Charger for AA, C, and D Batteries

Figure 7-20 shows a NiCad battery charger which recharges a NiCad in seven to fourteen hours, depending on initial charge still in the cell. The design allows you to charge two batteries of different sizes at the same time or four of one type at once. This unit is sold under the Archer name as Radio Shack No. 23-122.

Fig. 7-17. Special purpose batteries with voltage from 3 to 22.5 volts.

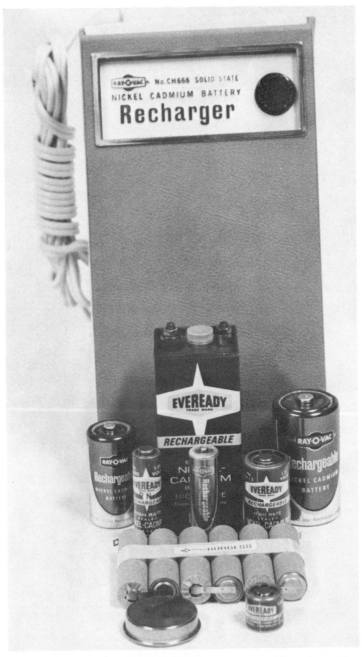

Fig. 7-18. Nickel cadmium battery charger with 115-Vac cord and representative rechargeable cells and batteries.

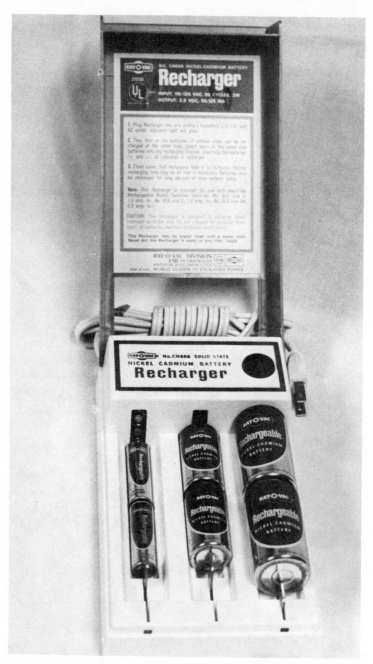

Fig. 7-19. Three different types of rechargeable cells being recharged at the same time.

168

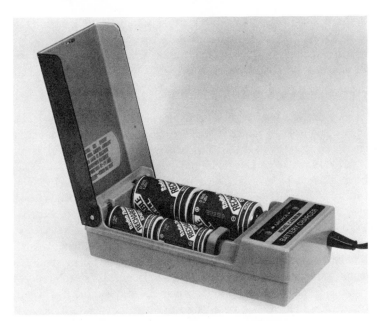

Fig. 7-20. Battery charger for NiCad type AA, C, or D-cells (courtesy Radio Shack, division of Tandy Corporation).

General Purpose Battery Charger

The battery charger shown in Fig. 7-21 is designed to recharge two or four regular or extralife batteries in sizes AA, C, D, or the

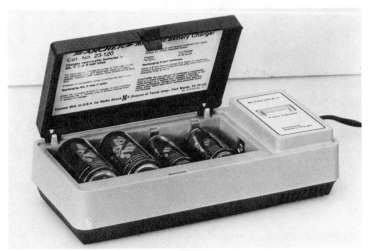

Fig. 7-21. General Purpose Battery Charger from Archer recharges up to four cells at once (courtesy Radio Shack, division of Tandy Corporation).

9-volt transistor battery. The charger automatically controls the charge rate so that batteries cannot be overcharged. It is good for recharging radio and flashlight batteries. This unit is sold under the Archer name as Radio Shack No. 23-120.

TESTER ACCESSORIES

Cell and battery testers assess the button cell used in watches and calculators and higher voltage batteries. With the advent of the transistor, battery operation has largely shifted to the 9-volt transistor battery or the use of multiple cells to provide 3, 4, 5, or 6 Vdc. Four or five decades ago, radios operated off the 90-volt to 135-volt battery which provided the high voltage necessary for vacuum tube operation in radios and amplifiers. Remember—at that time there were no portable radios, TV sets, audio tape recorders, video tape recorders, pocket calculators, smoke detectors, electronic quartz watches, computers, intrusion detectors, automatic door openers, lamp dimmers, automatic night lights, hearing aids, stereo receivers, electronic organs, personal pocket pagers, and the like. Electronics has indeed, "Come a long way, baby!"

Representative Battery Testers

In Figure 7-22 we see a tester which is a voltmeter with an adjustible switch set to the voltage to be tested. The scale is labelled to indicate whether the battery is good, weak, or needs to be replaced. The test leads are placed across the terminals of the battery to be tested according to the polarity marked on the battery.

In Fig. 7-23, we see a battery tester which is able to place a resistance load across the battery so that it is essentially tested "under load." In the figure, the Mallory battery tester switch is set to the 1.5-volt range and the load is set to the light drain. The light drain setting is used for testing batteries used in cameras, electric eyes, clocks, watches, calculators, and hearing aids. The medium drain setting is used for transistor radios, tape recorders, and other electronic equipment. The heavy drain setting is used for flashlights, lanterns, photoflash, and motor drives such as tape recorders.

A 1.5-volt battery is being tested in Fig. 7-24, with the instrument's probes placed across the battery terminals. The reading is in the Good portion of the scale, with a heavy drain setting.

Sensamatic Battery Tester

You can test all cells, including the 1.5-volt AAA, AA, C, and

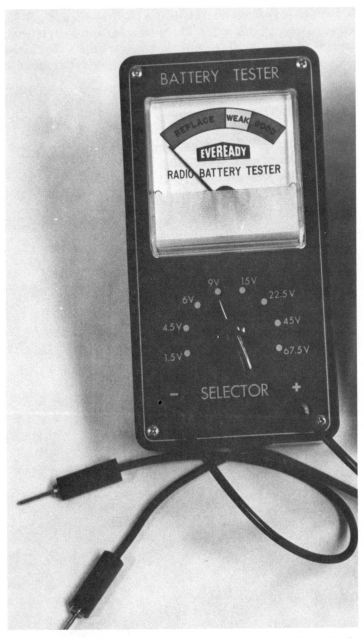

Fig. 7-22. This battery tester is a voltmeter with an adjustable switch set to the voltage to be tested.

D, by inserting them in this easy-to-use tester shown in Fig. 7-25, where a D cell is being tested. The Sensamatic will also test 6-volt lantern batteries as well as the 9-volt transistor battery. To do this, place the battery terminals across the test instrument as shown in

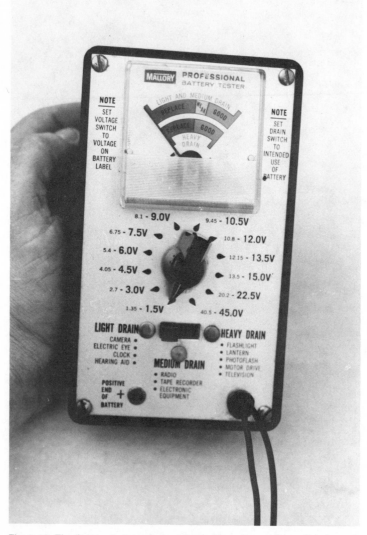

Fig. 7-23. The battery tester selects not only the battery voltage, but also puts varying loads on the battery to simulate actual in-use conditions. Battery tester set on 1.5-volt range with light drain switch setting.

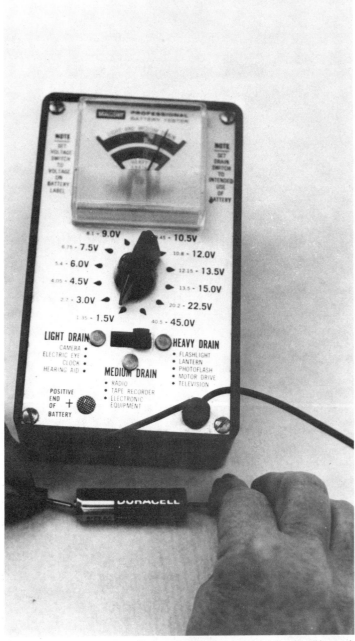

Fig. 7-24. Battery tester with heavy drain setting shows the 1.5-volt cell is good.

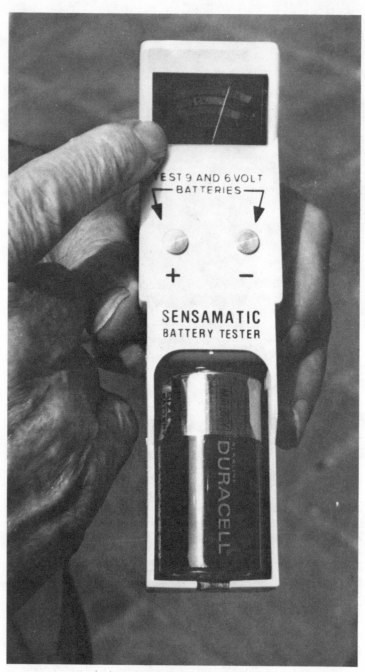

Fig. 7-25. D cell being tested in Sensamatic battery tester.

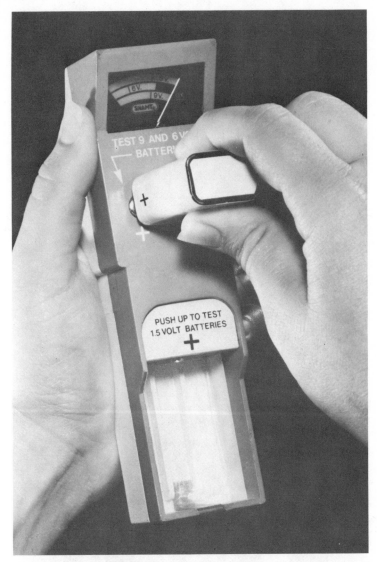

Fig. 7-26. 9-volt transistor battery tests good on Sensamatic tester (courtesy Cable Electric Products, Incorporated).

Fig. 7-26. This tester can be used with carbon-zinc, alkaline, and NiCad batteries.

Micronta Battery Tester

This battery tester has leads for testing nonstandard batteries.

As shown in Fig. 7-27, the meter will test batteries from small, light duty 1.5-volt button cell, to standard 1.5-volt cells, 6, 9, 12, and 22.5-volt batteries. The two color-coded scales are provided for regular and mercury batteries. The unit will also test NiCads, watch, and calculator batteries. The unit is available from Radio Shack as No. 22-030.

Micronta Battery Checker

The Micronta Battery Checker is small and handy to use on the

Fig. 7-27. Test meter with leads will test from 1.5 to 22.5-volt batteries (courtesy Radio Shack, division of Tandy Corporation).

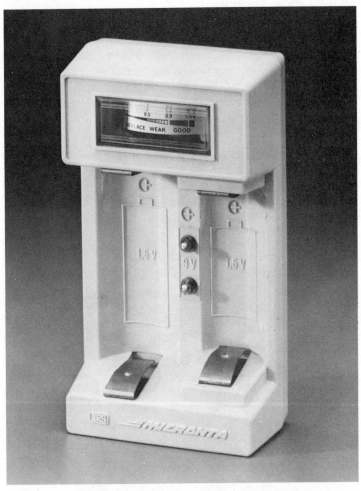

Fig. 7-28. This battery checker by Micronta will check sizes AA, C, D cells and 9-volt transistor batteries (courtesy Radio Shack, division of Tandy Corporation).

work bench or keep in your tool box. Shown in Fig. 7-28, the checker has a meter that indicates Replace, Weak, or Good, as well as a voltage scale for 1.5 and 9-volt cells and batteries. AA and C-cells are inserted on one side of the checker and D-cells are inserted in the other side. The 9-volt transistor battery terminals are placed across the two tabs in the middle of the tester. This unit checks carbon-zinc, alkaline, mercury, and NiCad batteries. This unit is available from Radio Shack as No. 22-100.

Micronta Button-Cell Tester

This button-cell tester is made especially for testing cells used in watches, calculators, cameras, hearing aids, and the like. Figure 7-29 shows the cell tester when the button cells are inserted according to the insert diagram on the front. The outer case of the button cell is positive and this arrangement makes it easy to hold the tiny cell to test. After the cell is inserted, a side-mounted switch is squeezed to get an indication of cell voltage. The unit will test silver oxide, alkaline, and mercury cells and the meter scale is marked to

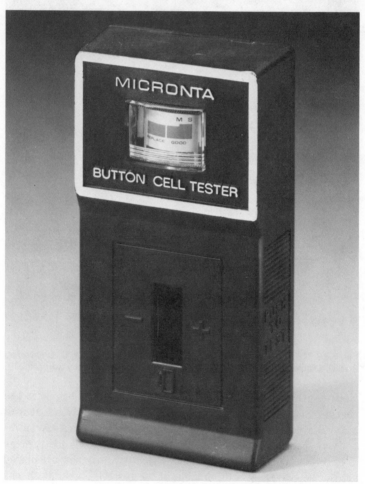

Fig. 7-29. The Micronta Button Cell Tester checks silver oxide, alkaline, and mercury cells (courtesy Radio Shack, a division of Tandy Corporation).

Fig. 7-30. 3-D battery checker by Micronta tests 9-volt batteries and 1.5-volt AA, C, and D cells (courtesy Radio Shack, division of Tandy Corporation).

show Replace or Good. An M or S on the scale is to be used for the mercury or silver cell. This handy, inexpensive tester is available from Radio Shack as No. 22-097.

Micronta 3-D Battery Checker

This battery checker is made to look like the batteries it tests. Figure 7-30 shows the checker with a 9-volt battery and a 1.5-volt cell with a tab for connecting the battery terminals. The unit will

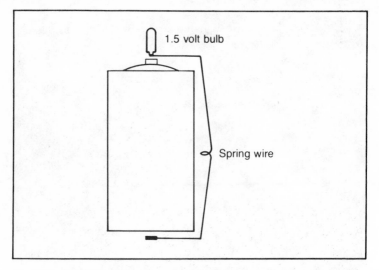

Fig. 7-31. A flashlight bulb and spring wire make a very handy cell tester.

test the 9-volt transistor battery as well as the 1.5-volt AA, C, and D-cells. A negative lead is connected to the bottom of the 1.5-volt cell when the positive terminal is placed on the top of the battery on the tester. Nine-volt batteries are tested by placing them directly across the tabs on the unit. A meter then indicates "Bad" or "Good." This tester is available from Radio Shack as No. 22-098.

FLASHLIGHT BULB CELL TESTER YOU BUILD YOURSELF

A very simple but most useful cell tester can be made by using a 1.5-volt bulb and a piece of spring wire as shown in Fig. 7-31. Since the bulb rating matches the nominal 1.5 volts of the battery, the brilliance of the bulb will be a very good indication of the battery strength. If the bulb is dim, the cell is weak and should be re-charged. If the bulb is bright, the cell is fully charged and can be placed into service. Be sure to use this tester only on 1.5-volts cells as it will not last very long when testing a 9-volt transistor battery!

Vehicle
8 Accessories

There are a number of vehicle accessories available for use to make your traveling, camping, hunting, and business trips safer and more enjoyable. In this chapter, we will cover some of those accessories.

GENERAL AUTOMOBILE STARTING TROUBLES

We all like to see spring and summer come around because the temperature begins to rise and we know that starting the car or truck in the morning will not be as much of a problem as in the winter. In general, we can say that starting a car or truck in cold winter weather is much more difficult than in summer.

What does cause most car-starting or truck-starting troubles? During the winter, most people blame the battery. The American Automobile Association places the blame of most winter starting problems on battery or electrical failure. The AAA elaborates on this by saying that cold-weather starting problems can usually be traced to three main items. These are battery deficiency, charging-system failure, and ignition trouble.

Battery Deficiency

How do we get into a battery deficiency situation when starting a car in winter? If your battery is old and needs replacing, you will notice it on the first cold morning when you try to start your car. The engine and its oil are cold and its like cranking the car with cold syrup for oil. When the car will not start, you will crank some more and shortly the battery is dead. Just plain pooped out! You will need a jump start to get the car going.

Perhaps you left your headlights on for several hours when you got to work. The battery would then be weak when you try to start the car. Or perhaps you might have driven at night with the headlights on and did not put enough energy back into the battery that night. You will know that the next morning when your car growls a few times and the battery goes dead. Again, you will need a jump-start—and remember, a battery that is going dead will not accept a good charge. You will just be postponing your problems if you do not replace the battery. But first, be sure to check the water level if you can, or the recharge capability if you have a sealed battery. To prevent premature battery corrosion, be sure the battery case and the area around the terminals are as free of moisture, dirt, road grime, and grease as is possible. Clean the terminals with a cleaning solution of water and ammonia or baking soda. Clean terminals can be coated with petroleum jelly. Be sure the battery cables are tight and replace any severely corroded cable terminals.

Charging System Failure

Battery failure is often a symptom rather than the cause of the trouble. The battery runs down in futile attempts to start the car. In some cases, the cause of a run-down battery is a loose alternator belt, not tight enough to drive the alternator. Watch your "idiot light" or ammeter to see if it remains on too long or too often, especially with the headlights or air conditioning on. A loose belt would mean the alternator is not putting enough current (energy) back into the battery, even if the car is brand new. With the engine off, push down on the alternator belt and you should get only about a half-inch of play. Adjust the belt as required. The charging system should be checked annually, particularly before the cold season.

Ignition Trouble

Statistics show that people who have their cars tuned up after experiencing their first starting trouble are likely to have fewer subsequent starting failures than those who simply buy a new battery. This good fortune will likely last the whole winter.

When you know that the battery is new and the alternator belt is properly adjusted, if the car still will not start, it can probably be traced to ignition or carburetion problems. If you have starting problems in summer, it obviously has to be one of these. A tune-up will usually include a check of the ignition harness as well as spark plug replacement, points, condenser, and the like. Replacing the ignition harness will give your car new starting zip as you are not losing the high voltage through leakage of the old, brittle wires. You have to spark to start!

Safety Tips on Auto Batteries

Some years after the automobile automatic shift was introduced, the now-familiar jumper cable was introduced. Figure 8-1 shows a pair of jumper cables that are used to "jump" from an auto's good battery to that of an auto with a "dead" battery. Dead enough to not start the engine, but perhaps alive enough to light the headlights, honk the horn and play the radio. The horn and radio do not take much power but the lights do, and the starter certainly does! That is true even in summer, when engine oil is oil and not slow-running jelly.

When the automatic shift first came out, drivers would give each other a start by pushing the car the way a manual shift auto can still be started. A manual shift, however, can usually be started just by pushing it by hand, shifting, and then letting the clutch out to turn the engine over. It is a fairly simple process. But to start an automatic shift required pushing the dead car until it was going 30 to 35 mph. The necessity for doing this was not apparent to all drivers, as one found out the hard way. He told the driver who was going to

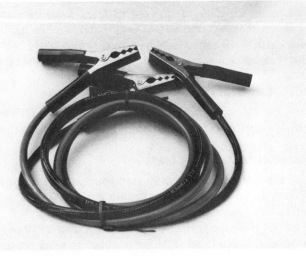

Fig. 8-1. Jumper cables for auto jump-starts. One cable is color coded Red (positive) and the other cable is colored Black (negative).

push him, "You will have to get going about 30 mph or so before my car will start." He got in his car and waited for the push. Then he looked into his rear view mirror and was horrified to see the driver coming at him doing about 30 mph!

Pushing a car to start is not recommended by auto manufacturers as this may damage the catalytic converter on gasoline engines or other parts of the auto. Jump-starting an auto can be simple but it does involve some safety considerations. Eye injuries caused by exploding auto batteries have risen to record highs as reported by the National Society to Prevent Blindness. As many as 15,000 people are treated each year for wet-cell battery-related injuries. Almost three-fourths of these injuries were to the eyes. Most of these injuries occur when jump-starting a car battery. To reduce the chances for an injury, people must be aware of the basics of battery maintenance and correct jump-starting procedures.

A Safety Precaution

When batteries are being charged, an explosive gas mixture containing hydrogen forms in each of the six cells of a 12-volt battery. Part of this gas escapes through the vent caps on the top of the battery and may form an explosive atmosphere around the battery. An explosion can result if an electrical spark occurs near the battery cell caps. A lighted match or cigarette, spark from a dropped tool, battery booster, or charge cables can ignite a battery. The resulting explosion can send a stream of sulfuric acid into the eyes, causing severe burns or even blindness. To minimize eye injury risk, each auto should have a pair of goggles at all times so that the driver can wear them while working around the car battery.

AUTO JUMP-STARTS

The following are procedures that should be followed when giving or getting a jump-start from one battery to another.

Procedure On Auto Jump-Starts

When you do have to jump-start your car or someone else's, to reduce the possibility of electrical or acid burns or an explosion, follow this procedure:

1. Move the vehicle with the booster battery close to the car to be jump-started so that the jump cables can reach but the cars do not touch. This assures that both batteries and their grounds are floating.

184

2. Turn off lights, heater, air conditioner, and all electrical loads of both cars. Set parking brake on both cars and automatic transmission lever of both cars in park. Manual shift cars ar placed in neutral. Turn off ignition of both cars.

3. Remove vent caps from both batteries, unless sealed.

4. Check discharged battery for proper fluid level and icing. If ice is present, remove battery from car and place in warm location until thawed. If necessary, add water.

5. Lay dry cloth over open vent holes of each battery.

6. Connect the red clamp (positive) of the cable to the positive (+) terminal of the booster (good) battery. Connect the other end of the red clamp (positive) to positive (+) terminal of the dead battery. This completes the plus-to-plus connection and is the safest.

7. Next, connect the black clamp (negative) cable to the negative (−) terminal of the booster (good) battery. Connect the other end of the black (negative) clamp to the engine block or a solid metallic ground in the car with the dead battery. This last connection is the most important one in jump-starts because you want to get this ground connection and any spark that might result at a safe distance from the dead battery.

8. Start the engine of the car with the booster battery. Let it run at fast idle for a few minutes so that it can put a quick charge into the dead battery.

9. Start the engine of the vehicle with the dead battery.

10. To complete a safe jump-start, reverse the cable connection sequence, removing the black (negative) clamp from the "dead" car's motor ground. Then disconnect the black clamps from the booster battery. Remove the black-to-black clamps first so that you will not have a plus and minus lead connected to either battery that could short out if you dropped one of the leads.

11. Re-install vent caps on both batteries and discard the cloths as they might have corrosive acid on them. Figure 8-2 shows the jumper cable connection sequence for jump-starts. Follow these procedures closely. Remember the sequence as, "Good RED to bad RED and good BLACK to bad GROUND."

Jump-Starting a Diesel Auto

When jump-starting a diesel engine vehicle that may have two batteries in parallel, make the boost connection to the battery that is closest to the starter. This electrical connection will be the shortest and have less electrical resistance. At low temperatures, it may not

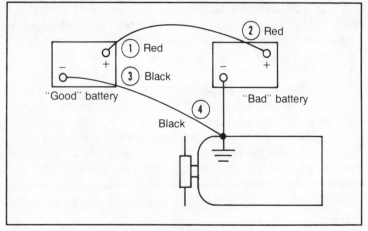

Fig. 8-2. Jumper cable connection sequence for jump-starts. Say, "Good RED to Bad RED and Good BLACK to Bad GROUND."

be possible to start a diesel from a single battery in another vehicle. When jump-starting from a service station or wrecker, be sure the equipment used is switched to 12-volt operation with negative ground. Do not allow the use of 24-volt charging equipment as this can cause serious damage to the electrical system or electronic parts in the diesel vehicle.

AUTO ACCESSORIES

In this section, some accessories will be described which can be used with your auto, truck, camper, or recreational vehicle.

Easy Charge by Permo

Easy Charge is a cigarette lighter-to-lighter car starter. It takes the old idea of jumper cables and refines it to a level of ease, convenience, and safety. Easy Charge does this by trading a quick jump-start for a time dependent slow charge of the dead or weak battery from a good battery. Figure 8-3 shows the Easy Charge cord and plugs. You just plug one end of the Easy Charge into the lighter of a "good" car, pass the 16-foot cord through both car windows and insert the other end in the cigarette lighter of the disabled car. A trickle charge immediately starts flowing into the dead battery. After a few minutes, the weak battery should be sufficiently re-charged to start the car. The unit comes in a handy plastic storage envelope small enough to fit in the glove compartment of any vehicle.

Fig. 8-3. The Easy Charge makes use of a cigarette lighter-to-lighter connection as a means of charging a vehicle battery (courtesy Permo).

Table 8-1 shows the charge time required for various engine starting conditions. Specific operating instructions for Easy Charge are as shown:

1. When car will not start, remove the cigarette lighter adapter plugs and connector from package.

2. Place the transmission in Park and engage the parking brake. Start the engine of the "good" vehicle. If inside a garage, open doors for ventilation! Run engine in well-ventilated area only.

3. Extend the wire cable through the windows to the vehicle which will not start.

Table 8-1. Easy Charge Condition Versus Time.

Condition	Time
Engine turns, but too slow	5 minutes or more
Engine will not run, but starter relay click heard when key turned to start	10 minutes or more
No click heard when key's turned to start	25 minutes or more

4. Remove the cigarette lighter in the good vehicle from its socket and insert the adapter plug into the cigarette lighter socket.

5. Turn on any interior light of the disabled vehicle such as dome light, map light, or glove compartment light.

6. Remove the cigarette lighter from the disabled vehicle. Push the cable adapter plug firmly into the lighter socket. Note the brightness of the interior light. If it becomes brighter when the adapter is plugged into the socket, good connection has been made and the disabled vehicle battery is being charged from the good vehicle. If no change in brightness occurs, reinsert one or both of the adapter plugs and rotate them slightly to establish a better electrical connection.

7. Charging time—It takes longer to charge a battery in temperatures below freezing (32°F). Refer to Table 8-1 for charging time. It will require less time in warmer temperatures.

Easy Charge is manufactured by Permo, 3001 Maimo Road, Arlington Heights, Illinois 60005.

Winco Inverter

An inverter is an electronic device that will take direct current from a battery and convert to alternating current voltage. The Winco Inverter shown in Fig. 8-4 converts dc battery power to 120-Vac power. These inverters are ideal for supplying 120 volts of power for agricultural, utility, industrial, marine, and recreational

Fig. 8-4. An inverter produces 120-Vac power from a battery supply of 12 V or 24 Vdc (courtesy Dyna Technology, Incorporated).

Table 8-2. Winco Inverter Models Versus Power Output.

Model	Watts	Input, volts	Output, volts
08FM2GC-91A	800	12 Vdc	120 Vac
102FM2GC-91A	1200	24 Vdc	120 Vac

applications. Lightweight and compact in design, the Winco Inverter has an automatic starting feature when the ac power is required and stops automatically when the ac load is removed. This eliminates unnecessary battery drain when 120 Vac power is not required.

By simply hooking up an inverter to a 12 or 24-volt battery, 120 volts of ac power is available for operating lights, drills, saws, routers, sanders, chain saws, and furnace motors.

The inverter is two machines in one: a dc motor (operating off the battery) and an ac generator. The motor and generator windings are wound on the same shaft. When power is applied to the dc motor, the inverter operates the ac generator on the same shaft, producing alternating current at the same time.

Two models are available from Winco. Table 8-2 describes the two models and their power capabilities. The 800-watt unit operates off 12 Vdc while the 1200-watt unit operates off 24 Vdc. The size of the battery will determine the length of time the inverter will operate. Batteries should be discharged no less than 50 percent of their capacity. That is, a 100-ampere battery should not be discharged over 50 amperes. Table 8-3 shows the average length of time the inverter will operate.

Figure 8-5 shows the Winco 800-watt model mounted permanently in a truck engine compartment. The unit is used to provide ac power for electric drills or spray gun use on a cherry-picker crane.

The Winco Inverters are manufactured by Dyna Technology,

Table 8-3. Battery Capacity Versus Operating Time.

	Battery Amperes Capacity					
	70	100	150	200	250	300
Full Load (Minutes)	21	30	845	60	75	90
Half Load	42	60	90	120	150	180

Fig. 8-5. Winco 800-watt inverter mounted in engine compartment of utility truck (courtesy Winco).

Incorporated, 7850 Metro Parkway, Minneapolis, Minnesota 55420.

Homemade Inverter for Car or Boat

The inverter described here is a lower cost version of the one described above. With this portable transistorized inverter, you can produce 115 Vac from a 12-volt car, boat, or camper battery. While the power is limited to 100 watts, it is enough to run many small electrical devices you normally enjoy only at home. These include a stereo phonograph amplifier, radio receiver, tape recorder, electric shaver, lamp bulbs up to 100 watts total, and even a small TV set.

The parts all fit in a 6 × 5 × 4-inch aluminum mini-box with a carrying handle on top as shown in Fig. 8-6. A flush type receptacle in the end along with a pilot light and an on/off switch completes outside parts mounting. You can plug directly into this outlet or connect an extension cord if you want power at some remote location, such as on the tailgate in a station wagon or inside a tent near the car. When not in use, the inverter should be unplugged and stored away.

The inverter is designed for 12-volt negative ground, the most common in use today. It is very important to observe correct polarity or the power transistors will be damaged. For quick, easy hook-up, the input leads can be wired to a handy cigarette lighter plug available at local electronics or auto stores. The plug then can be inserted in the lighter socket on the dashboard whenever you want ac power. For maximum efficiency, however, it is best to connect the positive lead directly to the positive terminal on the battery and the negative terminal to a ground connection on the car or boat. In either case, use heavy 12-gauge stranded wire as the leads must be capable of handling a hefty current flow. If the wires are thin or the connections not tight, output at the ac end will be reduced.

Referring to Fig. 8-7, we see the heart of the inverter is the step-up transformer, T-1. While the transformer is the most expensive item in the inverter, the finished unit will still cost you considerably less than commercial units of equivalent power capacity. The transformer and most of the other parts are available at local electronics supply stores. The center tap of the transformer (black lead) goes to the 12-volt positive (+) of the battery supply. The two halves of the primary are connected to identical transistor circuits. The transistors conduct alternately, producing a current flow first in one half, then in the other half of the primary. This action, in turn, creates a stepped up alternating current in the secondary. Since the transistors turn on and off 60 times a second, the current in the secondary has a frequency of 60 cycles per second (Hz) to match that of regular house current.

The transistors are high-power germanium pnp types as specified in the parts list in Table 8-4. They have these minimum

Fig. 8-6. This 100-watt inverter plugs into a car lighter and provides 115-volt 60 Hz power for use with low-power devices such as TV sets and shavers.

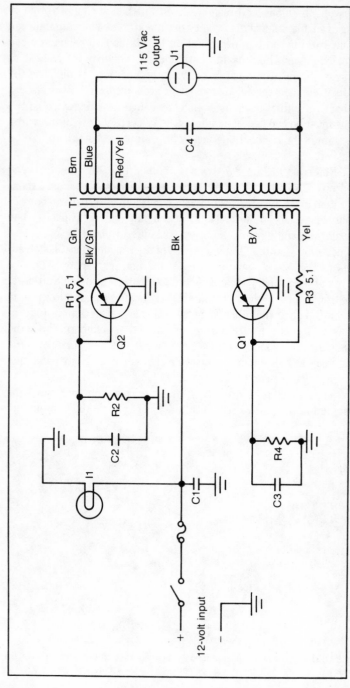

Fig. 8-7. Schematic diagram for 100-watt ac inverter. T-1 is a step-up- transformer.

Table 8-4. Parts List for 100-Watt Inverter.

Item	Description
C1	250 μF, 25 Vdc capacitor
C2, C3	2 μF, 50-Vdc capacitor
C4	1 μF, 400 Vdc capacitor
R1, R3	5.1 ohm, 5-watt resistor
R2, R4	200 ohm, 20-watt resistor
T1	Triad TY-75A step-up transformer, or equiv.
I1	12-volt indicator lamp assembly
J1	ac panel-mount outlet
Q1, Q2	High-power germanium PNP transistor (Motorola HEP 236, or alternates 2N3637, 2N3638, 2N3639)
S1	Heavy-duty SPST toggle switch with minimum 10-ampere rating
F1	Fuse holder. Use 10-ampere, 3AG fuse
Case	6 × 5 × 4-inch aluminum minibox

specifications: breakdown voltage of 36 volts, Beta of 50, and power rating of 150 watts.

The transformer is bolted to the end of the box, but rests on the bottom to carry its weight. The transistors are mounted on the outside back of the cabinet for good ventilation and cooling. Because they handle large currents and get warm, they must be fastened firmly so the box acts as a good heatsink. In connecting the transistors, note the markings on the leads to determine which is which. Most large power transistors of the type used here are stamped with a B or E to indicate base and emitter. The metal case of the transistor serves as the collector. If the leads should not be marked, hold the transistor so the two wires are horizontal and slightly above the center line of the case. The one on the left is the base and the one on the right is the emitter.

You will find several output taps on the transformer offering a choice of voltages. Select the one that gives the closest to the desired voltage. Start with the blue lead and check the output with a voltmeter. If the voltage is low, disconnect the blue lead and try the brown one. If the voltage is high, use the red/yellow tap. If a voltmeter is not available, compare the intensity of a 100-watt light bulb, first on regular house current, then on the inverter. It should have about the same brightness on both. If it seems too dim, switch to the higher-voltage tap; if too bright, the lower-voltage tap. Use lug-type terminal strips for making connections between components. Do not operate the inverter without a load plugged into the outlet.

Micronta Auto Battery and Charging System Analyzer

An extremely handy device that uses solid-state technology to tell you some specifics about your car battery and charging system is the Micronta Auto Battery and Charging System Analyzer. Shown in Fig. 8-8, the unit is plugged into your cigarette lighter socket and goes to work immediately to tell the good and bad aspects about your auto's electrical system. You can do this before you are stranded on the road someplace with a dead or damaged battery.

The unit is completely solid-state, with an LED readout (red, yellow, green) that helps pinpoint the problem for faster troubleshooting. The analyzer is about the size of a half-pack of cigarettes and will store in the glove compartment. The analyzer is meant to be used on 12-volt negative ground vehicles.

Checking the Battery. To check the condition of your vehicle battery, follow these steps:

1. Plug analyzer into lighter socket.

2. With engine off, turn headlights on low beam (this puts a drain on the battery).

3. After one minute, read the LEDs (with engine off).

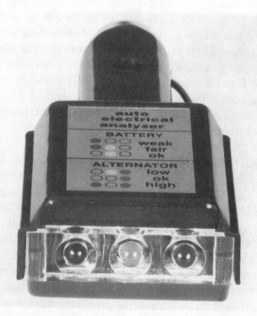

Fig. 8-8. Micronta Auto Battery and Charging System Analyzer plugs into lighter socket (courtesy Radio Shack, division of Tandy Corporation).

Table 8-5. Battery Analyzer Indications.

Reading	Indication
RED	Low battery charge, battery damage, low voltage regulator setting, loose alternator belt, faulty wiring, damaged alternator diodes or slip rings, idle adjustment too low
RED and YELLOW	Battery is in fair condition. Electrolyte level may be low
YELLOW	Battery condition is good

After you have tested your battery, its condition can be assessed by referring to Table 8-5.

Checking the Alternator and Regulator. To check the condition of your alternator, belt adjustment, and regulator, follow these steps:

1. Run engine at fast idle, or drive car for 15 minutes (this places a charge in the battery).

2. Read the LEDs.

After you have run your engine for the nominal 15 minutes, and have placed a charge from the alternator into the battery, use Table 8-6 to ascertain the conditions as shown. After completing all tests, remove the analyzer from the lighter socket as it draws a small amount of current. On long trips, however, it can be left plugged in to help advise you of the condition of your electrical system. If something is wrong, the analyzer will spot it.

The Auto Battery and Charging System Analyzer is marketed under the Micronta tradename by Radio Shack as Catalog No. 22-1635.

Table 8-6. Alternator and Regulator Analyzer Indications.

Reading	Indication
YELLOW and GREEN	Insufficient charging due to excessive current consumption (short in system); misadjusted or defective voltage regulator
GREEN	Charging system operation normal
RED and GREEN	Battery is being overcharged. Causes include misadjusted regulator, poor ground or other poor wiring (especially in regulator circuit), or defective regulator

Fig. 8-9. The Carlite by Black and Decker operates off 12-volt cigarette lighter socket. Red flasher lens cover is on left (courtesy Black and Decker).

Carlite by Black and Decker

The Carlite is an ideal auxiliary light for cars, trucks, boats, or farm implements. It operates from most 12-volt cigarette lighter sockets. The beam width is designed for close illumination or penetrating long-distance use. Shown in Fig. 8-9, the Carlite has a built-in flasher and red lens to provide a reliable emergency signal. The red emergency lens cover is shown in the left of Fig. 8-9. The Carlite has a built-in hanger and a height adjuster lets you position the light to the desired work area. A cord wrap area is convenient for cord storage.

A front guard protects the bulb and keeps the beam heat away from close objects that might contact the lens. The Carlite weighs just 1½ pounds and has a two-year home warranty. The sealed beam lamp unit operates off 12.6 Vdc and draws 2.6 amperes. A 12-foot cord with a lighter socket adapter is provided with a fused plug. A three-position switch provides off, light, and flash capabilities.

The Carlite is manufactured by Black and Decker, Incorporated, Towson, Maryland 21204, and is available through local distributors.

Magnetic Utility Light

As shown in Fig. 8-10, the Magnetic Utility Light plugs into a car's lighter socket. A strong magnet attaches the light to your engine hood, trunk, or auto body so that you can perform emergency road repairs at night. A 12-foot retractable cord supplies power to the light.

The Magnetic Utility Light is sold by Radio Shack as Catalog No. 61-2502 and is available at your local stores.

Fig. 8-10. Magnetic Utility Light from Radio Shack operates off 12-volt lighter socket (courtesy Radio Shack, division of Tandy Corporation).

Flexible Map Light

This handy map light plugs into a lighter socket and is compact enough to store in your glove compartment. It has a flexible 5-inch long neck and an On/Off switch for convenient use when traveling at night and provides light for map reading, and the like. Figure 8-11 shows the unit which will provide concentrated light for use but will

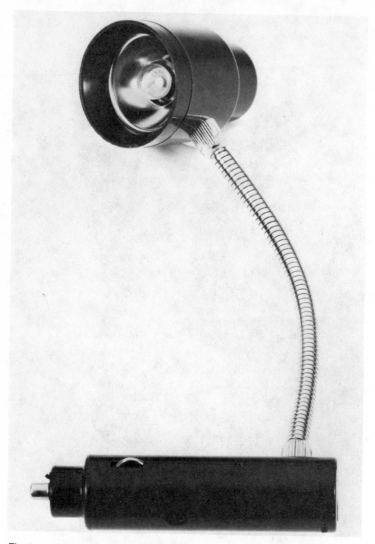

Fig. 8-11. Flexible Map Light by Radio Shack plugs into lighter socket (courtesy Radio Shack, division of Tandy Corporation).

Fig. 8-12. The Archer Headlight Alert reminds you to turn off your lights when you park (courtesy Radio Shack, division of Tandy Corporation).

not bother the driver's night vision.

The Flexible Map Light is sold by Radio Shack as Catalog No. 61-2501.

Auto Headlight Alert

This little Auto Headlight Alert will remind you to turn off your headlights when you park your car. While turning off headlights is a simple thing to do, sometimes you will lock your car and walk off without doing it! Since your car is locked, others that see cannot help. Hours later, you will come back and listen to your starter growl at you when it will not start. Then you will growl at yourself!

Figure 8-12 shows the Headlight Alert which will remind you to turn your headlights off by beeping at you for 10 seconds when you leave your lights on and the engine is off. Then, a red LED flashes until the lights *are* turned off. The alert is complete with self-tapping screws and double-sided tape. It installs easily in most cars with accessible fuse blocks in just a short while. It may save you a ride in a tow truck!

The Auto Headlight Alert is marketed under the Archer logo by Radio Shack as Catalog No. 270-110.

9 | Solar, Wind, and Manpower Generators

The sun is the source of all energy that we have on earth. Solar energy we receive now, but other forms of solar energy were placed here on earth eons ago. We receive light and heat energy eight minutes after it leaves the sun. But energy from coal and oil takes a bit longer. Millions of years ago, giant ferns and similar plants lived and died on earth. When they were alive, they trapped the sun's energy and manufactured food which they stored in their tissues. When the plants died, they sank down into swampy soil. After millions of years, they were changed into coal, which we now dig up. When we burn coal, it releases the light and radiant heat of the sun which the plants trapped many millions of years ago.

Petroleum, or oil, was probably formed in the same manner. Scientists believe that eons ago, many jillions of tiny plants died and their tissues were changed into oil deep in the ground. So when we pump oil out of the ground and use it to make fuel, we are releasing the trapped energy of sunlight which shone on earth millions of years ago. Both coal and oil are really "sunpower."

The energy which we as humans have is really due to solar energy as well, because there would be no food to eat without the sun. Plants grow, using the sun's energy, and we eat the plants. Animals eat the plants and change that energy into body tissue. Since we eat both plants and animals, we receive energy from both. The energy in both plant and animal food came first from the sun. So when you eat food, you are eating "stored sunlight."

So far we have discussed how fossil fuel and manpower energy are derived from the sun. But what about wind power? Well, it also can be called "solar power" due to the action of the sun which makes the wind. The sun also lifts millions of tons of water into the air and makes rain when it falls to earth, which flows into rivers to power giant electrical turbines at dams. Essentially, all energy on earth is traceable to the sun. In this chapter, we will discuss aspects of solar, wind, and manpower.

SOLAR POWER

About 1000 watts of solar power per square yard are received on earth at the equator. That is about 1½ horsepower per square yard! Lesser amounts of power are received at other parts of the earth. All we have to do is figure out a good way to extract and use this energy. Solar water heaters have become popular in the past decade, since the oil crisis. That energy crisis made everyone more aware of energy demands, and pointed out our need to find old and new ways of conserving energy. One way of conserving our present stored fuel is to use the solar energy from the sun in the form of radiant energy.

Solar or Photovoltaic Cell

The solar cell is one of the most practical means of converting the sun's radiant energy directly into a useful product—electricity. At the present time, solar cells are about 10 percent efficient in extracting radiant energy from the one square yard which receives the 1000 watts of power. This means that one square yard of photovoltaic cells would produce 100 watts of electrical power. Because of the cost of solar cells, a decade ago it cost $500 to produce one peak watt. Today, the cost for solar cell power is less than $10 per peak watt. With a tax incentive energy payback, the cells will pay for themselves in about three and half years. Both the cost and energy payback are continuing to decrease as popular usage and research drives them down. By 1985, costs will drop to $4 per watt and energy payback will be reduced to less than one year.

A silicon photovoltaic cell is illustrated in Fig. 9-1. A typical cell consists of a two-layer silicon crystal. The N(negative) layer contains an impurity (it has been doped) which provides negatively charged free electrons. The P (positive) layer also contains an added impurity different from the dope used in the N-layer. The P-layer produces free roaming positive charges (called "holes"

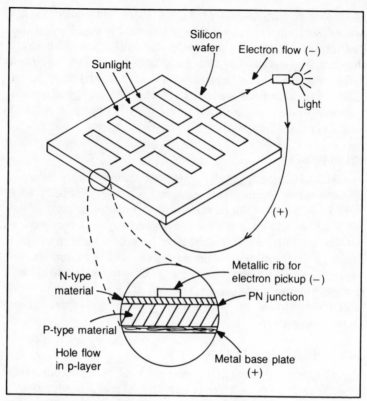

Fig. 9-1. Typical photovoltaic cell consists of P-layer and N-layer silicon crystal. Light is converted directly into electricity with about 10 to 12 percent efficiency.

because they lack an electron).

When sunlight strikes the N-layer of the silicon cell, these opposite charges (electrons and holes) diffuse across the P-N junction and an electric current is generated through the circuit completed by the light bulb. Note that the metal base plate of the cell is positive, while the metal pickup wire, or rib, is negative. Clear, chemical coatings provide protection to the solar cell.

Solar-Cell Materials. The first barrier-layer photovoltaic cell was developed in about 1876 using selinium. Several materials have been used in the fabrication of photovoltaic devices since 1876, primarily cadmium sulfide, silicon, and gallium arsenide. Silicon is widely used since it is the second most abundant element on earth, is low in cost when used in low purity form, and provides high photovoltaic efficiency.

Ongoing Solar-Cell Projects. The number of projects in the country using solar energy to generate electrical power increases every day. Solarex has an industrial structure powered completely from the sun. The roof of the building is slanted to the south and has a 28,000-square foot roof covered with 3000 semicrystalline panels. These panels provide about 200 kW of photovoltaic power for the facility's production lines which produce photovoltaic cells and panels. The panel array produces about 800 kWh per day of energy which is stored in a 2.5 MWh battery-storage system. The storage system provides 60 kW of uninterrupted, steady power should four dark days occur in succession.

Arco Solar is building a new solar plant project which will produce 6.5 MW of photovoltaic power from a plant located near San Luis Obispo, California. It is expected to produce enough power to supply 2,000 homes. The plant will sell its electricity to the Pacific Gas and Electric Utility company. The solar plant will eventually increase its peak power generating capability from 6.5 MW to 16.5 MW.

Portable Solar Electric System

A solar electric system that can be transported in its own easy-to-carry case is shown in Fig. 9-2. Each portable unit contains

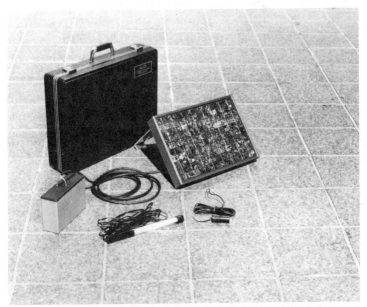

Fig. 9-2. Portable solar electric system comes complete with carrying case. Energy is stored in battery (courtesy Solar Contractors, Incorporated).

a solar module, a battery, a 12-volt socket, a light that provides 40-watts of illumination and a battery tester. One model contains a 120-volt, 100-watt dc to ac inverter. Because of their small size and weight, these units can be carried any place where electric power is needed. They are a renewable electric energy source and are maintenance-free as they have no mechanical parts.

The portable systems are available in three models. The SCP 8 has an 8-ampere battery capacity at 12 Vdc; Model SCP 12 has 12 amperes of battery capacity at 12 Vdc and Model SCP 100 ac will provide 12 amperes of battery capacity and 100 watts of ac power at 120 volts. The case for the SCP 8 model measures 20 × 14 × 7 inches while the case for the other models measures 24 × 18 × 7 inches. The weights for the three models vary from 22½ to 36 pounds.

These portable solar systems are available from Solar Contractors, Inc., 8 Charles Plaza, #805, Baltimore, Maryland 21201.

Locally Available Solar Power Devices

There are several solar panels and cells available at your local Radio Shack stores. You can experiment with them to learn how solar cells work, what you can do with them, and perhaps decide that you want a larger installation to meet your needs at home, office, or plant. All of these devices are fairly inexpensive and they will give you a start in the energy supplier of the future—the solar cell.

Silicon Solar Cell. This is a single silicon solar cell that measures 1 × 1.6 inches and produces approximately 0.2 amperes of current at 0.42 Vdc in full sunlight. This voltage is the basic building block and you can use several of them in series to charge batteries or to power small projects. This is Radio Shack Catalog No. 276-124 and costs about four dollars. It is available at local stores. Figure 9-3 pictures the cell.

Solar Energy Project Kit. This is a silicon solar cell pre-wired to a precision dc motor. The unit shown in Fig. 9-4 is perfect for building model windmills, boats, science fair projects, and the like. The kit comes complete with a minipropeller and project book. Radio Shack carries this unique kit as Catalog No. 277-1201.

Portable Solar Panel. For those who would like to get serious about solar energy and its uses, this portable solar panel is just the way to get started. Shown in Fig. 9-5, the solar panel has two detachable reflectors. The panel contains 32 first-quality silicon cells, which under bright sunlight, will convert the solar energy

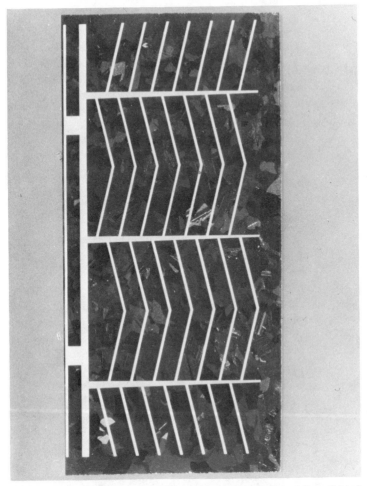

Fig. 9-3. This silicon solar cell will produce 0.42 Vdc under full sunlight. It measures 1 × 1.6 inches (courtesy Radio Shack, division of Tandy Corporation).

into approximately ½ watt of free electrical power. The solar panel is ideal for charging batteries or powering small devices directly. Its typical output is 80 mA at 6 Vdc or 40 mA at 12 Vdc. The panel is switchable from 12 Vdc to 6 Vdc. It is supplied with a 48-inch plug-in power lead with clips for easy connection to your project.

This portable solar panel measures 5¾ × 4⅛ × ⅝ inches. The special lens and detachable reflectors measure 5 × 4 inches each. The panel is not designed for permanent outdoor installation, however. The panel costs about $25 and is carried by Radio Shack as Catalog No. 277-1250.

Fig. 9-4. Prewired solar cell and motor kit will drive small propeller from energy in sunlight (courtesy Radio Shack, division of Tandy Corporation).

Fig. 9-5. This portable solar panel produces 6 or 12 Vdc at 40 mA in bright sunlight (courtesy Radio Shack, division of Tandy Corporation).

WIND POWER

The wind blows almost incessantly along the coast, and most of the time it is a nuisance to swimmers, bathers, fishermen, beach strollers, picnickers, and beach-front owners. But in some parts of the country, investors are paying large sums of money for land that no one could unload at any price a few years ago. The reason is strong winds, which may make the land worthless for farming, living, grazing, or recreation but are ideal for wind farming.

One area that has prospered from this gold blowing in the wind is California. Almost overnight it has turned into a wind-farming center. There are already over 1,000 wind turbines there generating electricity sold to the public utilities. California has a number of spots that have steady, strong winds funneling in over land that has only marginal economic value. These are areas such as the rock-strewn deserts in Riverside Country just north of Palm Springs, the barren mountain passes in the Tehachapi Mountains in southern California, and northern California's Altamont Pass, which is about 50 miles southeast of San Francisco.

There are many parts of the country where the winds are good above land that is unsuited for agriculture or residential development. Upper New York state has some areas as does Texas, especially along the Gulf coast. The Wind River area of Wyoming is also a windy spot and there are a few other spots in New England. In some parts of the country, however, the price of utility-provided electricity is not high enough to justify the heavy capital investment in wind-turbine generators and the associated high-tension lines that criss-cross the country.

MOD-5A Wind Turbine

The island of Oahu in Hawaii will soon receive a wind-turbine generator that will produce 7.3 MW of electrical power. The turbine is being built by General Electric and will be installed and operational on the windy North Shore of Oahu sometime in 1985. The turbine will generate 30 million kWh yearly, which is enough power to meet the electricity demands of about 3,500 homes.

The MOD-5A turbine will have a 400-foot two-blade rotor with a shaft connected to a nacelle mounted on a tower 240 feet above ground. The turbine will start operating when the wind reaches a speed of 14 mph. It will produce maximum power at a wind speed of 32 mph and will shut down automatically when the winds are higher than 32 mph. The whole structure is designed to withstand hurricane winds of 130 mph. A special gear drive will step up the rotor speed of 16.8 rpm to 1,200 rpm in order to turn the electric generator. The whole wind turbine will weigh about 125,000 pounds.

The blades of the giant rotor will be built of wood laminate to minimize television interference. Just as overhead passing aircraft can interfere with TV reception, especially those sets with rabbit ears for an antenna, the giant blades can also cause interference fading as they rotate slowly high in the sky.

Wincharger Wind-driven Generator

The wind-driven charger by Wincharger shown in Fig. 9-6 puts out 450 watts from a wind-driven generator. The generator puts out either 12 or 24 volts of dc power to operate lighting, appliances, and electrical equipment. A 200-watt model is also available, with a 6-foot propeller. The 450-watt model has an 8-foot propeller and is designed for easy installation of the unit and tower. The governor is preassembled at the factory. In winds of approximately 20 mph, the 450-watt directdrive Wincharger will provide 30 amperes of current. When connected to a bank of batteries, the Wincharger provides electrical power for residential and marine applications.

Fig. 9-6. This wind-driven generator has an 8-foot prop and produces 450 watts of power when wind speed is above 20 mph (courtesy Dyna Technology, Incorporated).

A new improvement on the wind-driven generator is an ac current rectified to dc current. The generator has fewer brushes and, therefore, requires less maintenance. The propeller on the 450-watt Wincharger is constructed of laminated hardwoods for greater strength and longer life. A solid-state voltage regulator is also available.

An ac inverter described earlier in Chapter 8, will convert the 12 or 24 Vdc power to 115 Vac power. The user can combine a Wincharger and a battery bank to build a self-contained, independent 115-volt electrical system that is especially useful at a hunting lodge, retreat, or on the farm. Such an installation would also be helpful for radio amateurs during their operations for the annual field-day or contest operation. Priced at about $1000, the Wincharger wind-driven generator is manufactured by Dyna Technology, Incorporated, 7850 Metro Parkway, Minneapolis, Minnesota 55420.

MANPOWER GENERATORS

Man's research in other fields has progressed so well that he has not had to devote much thought to man-powered generators. With the invention of the transistor bringing low-power consumption, satellite communications rescue for downed flyers, giant strides in battery development, and pocket pagers for the pizza delivery boy, there has been little need to develop body-powered electrical devices. Therefore, man-powered generators have taken a back seat in the technologically advancing field of power generation. No longer do you see movies of the army GI pedaling or arm-cranking a field generator as was so popular during World War II. Even the popular "Gibson Girl" hand-cranked rescue radio transmitter has been forgotten. Only in black-and-white movies will that technology be remembered. As the state of the art of power generation advances, there is little literature available on man-powered generators. Other modes of power generation are too far advanced to spend research and development dollars on hand-cranking or bike-pedaling generators.

A Man-Powered Airplane. A recent, history-making event having to do with man power is the flight of the Gossamar Albatross across the English Channel, which took two hours by man-powered pedaling. To find out how much energy that feat took, consider the following. During field-day contests in which radio amateurs venture out into the field to test their radio amateur transmitting and receiving equipment (and their chigger stamina!), extra contest points are awarded for those who power their equipment by means other than commercial mains or petroleum derivatives. These other means should include alternate energy sources such as solar, wind, methane, grain alcohol, and so forth. A bicycle-powered generator comes to mind as being a suitable alternate means of supplying electrical power.

To determine how much leg-power is required to develop enough horsepower to power a radio transmitter, let us consider the following. The transmitter consumes 3.1 amperes at 13.8 Vdc. This is equal to 43 watts, about 0.06 hp. When you are pedaling a bicycle-mounted generator, this is about the point where leg pressure becomes noticeable. At 5 amperes, which is about 0.1 hp, it is very noticeable. And at 10 amperes, it is almost impossible to

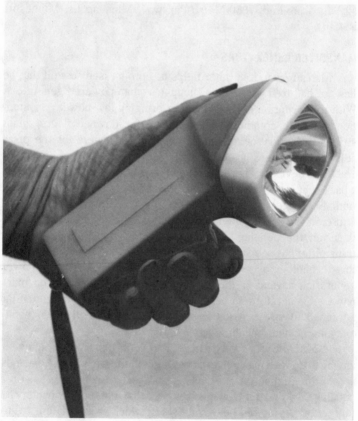

Fig. 9-7. Hand-cranked flashlight is powered by small dynamo by repeatedly squeezing the grip.

maintain this rate for an average middle-aged man for even a minute. But Mr. Bryan Allen, who pedaled the Gossamer Albatross across the English Channel, developed 0.9 hp for over two hours! He had to lift himself, the airplane, and the engine, so he developed the equivalent of 49 amperes of current! That is, 49 amperes times 13.8 volts is equal to 676.2 watts, or 0.906 horsepower (1 hp = 746 watts).

Hand-Cranked Flashlight

Let us get down to earth now and talk about a hand-cranked flashlight. Figure 9-7 shows a handcranked flashlight which produces light when you squeeze a grip to rotate a miniaturized generator to produce voltage. The unit can not rust, leak, or corrode—ever! You will never be disappointed by dead batteries. The hand-cranked flashlight is ideal for home, auto, garage, and camping. This unit can store easily in a glove compartment and is equipped with a strap to hang on a nail or to wrap around your hand for safe-keeping. You will always have electrical power for the light just by squeezing the grip. So you can take your exercise and be enlightened at the same time.

A dynamo flashlight is available as No. 8M10400 from BNF Enterprises, P.O. Box 3357, 119 Foster Street, Peabody, Massachusetts 01960, for under $10.

Index

Edited by Steven Bolt

About the Authors

Rudolf F. Graf and Calvin R. Graf are not related, have not yet met, and are friends by means of the telephone and mail. A common bond of technical writing brought them together for this, their first book as a team. Rudolf Graf has been in the electronics profession for more than 30 years and is a graduate in communications engineering from Polytechnic Institute of Brooklyn, and has an MBA from NYU. A radio amateur and senior member of the IEEE, he has written numerous books and articles on electronics, mechanics, and automotive engineering topics. He also taught electronics for several years.

Calvin Graf graduated from the University of Texas, Austin, and has been writing technical books and articles for over 30 years, covering antennas and propagation, radio reception, and circuit design. He served as a senior member of the IEEE for 25 years and was chairman of the Central Texas Section of the IEEE. He is an electronics engineer with the U.S. Air Force at Kelly AFB, San Antonio, Texas. He is a lecturer on the evening division faculty of San Antonio College, where he teaches classes in electronics theory and laboratory practices.

OTHER POPULAR TAB BOOKS OF INTEREST

TAB TAB BOOKS Inc.

Blue Ridge Summit, Pa. 17214

Send for FREE TAB Catalog describing over 750 current titles in print.